Corrosion Engineering

Corrosion Engineering

Editor

Ombir Singh Yadav

Corrosion Engineering

Edited by **Ombir Singh Yadav**

Printed in 2017

ISBN: 978-1-68117-350-4

Library of Congress Control Number: 2015939263

© 2016 by
SCITUS Academics LLC,
616, Corporate Way, Suite 2, 4766,
Valley Cottage, NY 10989

www.scitusacademics.com

Contents

Preface vii

Chapter 1   Efficiency of Amino Alcohols as Corrosion Inhibitors in
Reinforced Concrete 1
Ioannis Vyrides, Eleni Rakanta, Theodosia
Zafeiropoulou, and George Batis

Chapter 2   Greener Approach towards Corrosion Inhibition 21
Neha Patni, Shruti Agarwal, and Pallav Shah

Chapter 3   Study of Flow-assisted Corrosion of AZ91D
Magnesium Alloy in Loop System Based on Array
Electrode Technology 47
Hualiang Huang, Guoan Zhang, Jiakuan Yang,
Zhiquan Pan, and Xingpeng Guo

Chapter 4   Effect of Corrosion Inhibitors in Limestone Cement........................ 67
Evgenia Zacharopoulou, Aggeliki Zacharopoulou, Atteyeh
Sayedalhosseini, and George Batis,
Sotirios Tsivilis

Chapter 5   Effect of Residual Stress on the Corrosion Behavior of
Austenitic Stainless Steel.. 85
Osamu Takakuwa and Hitoshi Soyama

Chapter 6   Electrochemical Investigation of Corrosion on AISI 316
Stainless Steel and AISI 1010 Carbon Steel: Study of the
Behaviour of Imidazole and Benzimidazole as Corrosion
Inhibitors 109
Roberta R. Moreira, Thiago F. Soares,
and Josimar Ribeiro

Chapter 7   Corrosion and Surface Treatment of Magnesium Alloys................ 137
Henry Hu, Xueyuan Nie, and Yueyu Ma

Chapter 8   Corrosion Protection of Magnesium Alloys in
             Industrial Solutions 195
             Amany Mohamed Fekry

Chapter 9   Corrosion of Materials in Liquid Magnesium Alloys and
             Its Prevention 227
             Frank Czerwinski

             Citations...279
             Index...283

Preface

Corrosion Engineering is the specialist discipline of applying scientific knowledge, natural laws and physical resources in order to design and implement materials, structures, devices, systems and procedures to manage the natural phenomenon known as corrosion. Generally related to Metallurgy, Corrosion Engineering also relates to non-metallics including ceramics. Corrosion Engineers often manage other not-strictly-corrosion processes including (but not restricted to) cracking, brittle fracture, crazing, fretting, erosion and more. Corrosion engineering groups have formed around the world in order to prevent, slow and manage the effects of corrosion. Examples of such groups are the National Association of Corrosion Engineers (NACE) and the European Federation of Corrosion (EFC), see Corrosion societies. The corrosion engineers main task is to economically and safely manage the effects of corrosion on materials. Corrosion Engineering Masters degree courses are available worldwide and are concerned with the control and understanding of corrosion.

Editor

Chapter 1

Efficiency of Amino Alcohols as Corrosion Inhibitors in Reinforced Concrete

Ioannis Vyrides[1], Eleni Rakanta[2], Theodosia Zafeiropoulou[2], and George Batis[2]

[1]Department of Chemical Engineering and Chemical Technology, Imperial College, London, UK

[2]School of Chemical Engineering, National Technical University of Athens, Athens, Greece

ABSTRACT

The objective of this paper is to investigate the behaviour of amino alcohol corrosion inhibitors when they are used in reinforced cement mortars either as admixtures in the cement paste or as coating applications on the surface of the rebars. The reinforced cement mortars were exposed to both partial and full immersion in 3.5 wt% NaCl solution. Electrochemical measurements such as half-cell potential and linear polarization technique, as well as weight loss of the embedded rebars were performed in order to obtain information on the corrosion behaviour of the reinforcing steel in cement mortar. Results demonstrate that the amino alcohol corrosion inhibitors offer protection against rebar corrosion in cement mortars.

INTRODUCTION

Under most conditions concrete provides a reliable protection against corrosion of reinforcing steel, however corrosion remains the most common cause of deterioration of reinforced concrete. In order for corrosion to occur, the presence of an electrolyte such as the aqueous phase in concrete is required. The final result of the corrosion process is the formation of a thick layer of rust which exerts sufficient tensile forces within the concrete to cause cracking of the concrete cover [1]. In an alkaline solution, such as the calcium hydroxide solution in wet cement, a protective oxide film is formed which covers the steel thus passivates it. Corrosion occurs when this protective film is impaired and oxygen is present [2]. Oxygen has the most detrimental effect; chloride is only the catalyst of the corrosion process. Oxygen binds the electrons originally associated with the iron atoms and it is the abstraction of these electrons which allows some iron atoms to dissolve as ferrous ions and then to preipitate as ferric oxides [3]. Without oxygen steel rebar will not corrode in alkaline water even with chloride present [4]. Moreover, the stability of the film depends on the maintenance of a certain minimum pH value above which access of oxygen will not cause corrosion. However, if the access of carbon dioxide reduces the pH to a value 10 or lower the film is impaired, the natural passivity of concrete is thus reduced and under such conditions any access of oxygen will cause corrosion [5]. The presence of chloride ions stimulates corrosion by raising the pH required to stabilize the passive film to a value which may exceed that of a saturated calcium hydroxide solution. The intrusion of chloride ions depends on the porosity and permeability of the concrete material [1]. As the reinforcing steel corrodes, the ferric oxides occupy a volume three to four times greater than the initial reinforcing steel resulting in bursting stresses that crack and spall the concrete cover.

Corrosion inhibitors have been used successfully in steel pipelines, tanks, etc., for many decades. Their use as admixtures to concrete, however, is more recent and more limited. This reluctance is due to the fact that they could not be changed if found to be ineffective, replenished if found to be consumed, or removed if found to have deleterious effects [6]. The flexibility of corrosion inhibitors with regard to dosage and their compatibility with all aspects of construction and operation of structures makes them useful for protection against corrosion. Any inhibitor should have good solubility characteristics and rapidly saturate the corroding surface. Also the physical and durability properties of concrete should not be adversely affected [7].

Amino alcohol corrosion inhibitors control corrosion by attacking the cathodic activity, blocking sites where oxygen picks up electrons and is reduced to hydroxyl ion. Also inhibition of corrosion occurs through a mechanism whereby amino alcohols displaces chloride ion and forms a durable passivating film. In this view, although the amino alcohols adsorb on non-corroding sites which may seem more cathodic than anodic, they can just as easily be said to adsorb on potentially anodic sites as well [3]. Ormellese et al. [8] suggest that organic corrosion inhibitors reduce the ingress of chloride by filling concrete pores and blocking the porosity of concrete by the formation of complex compounds. Thus, the value of chlorides reaching to the steel surface is significantly less so the corrosion is inhibited. Several studies of the corrosion inhibition effect of amino alcohols for steel report their performance as a function of concentration and pH in saline solutions [9-11]. In particular, the colorimetric and electrochemical study of 2-ethylaminoethanol observed that the molecules do not provide a protective film in neutral media, but rather enhance the corrosion rate. At pH > 10.3 a protective layer of metalhydroxide is formed, which, however, is not completely insoluble. This hydroxide film can be stabilized by adsorption of 2-ethylaminoethanol on the hydroxylated surface and presumably by chelation of the ferrous ion [12,13].

An amino alcohol corrosion inhibitors can be mixed within the fresh concrete (admixture inhibitor), or can be applied on the surface of existing concrete structures. In this study the contribution of both types' admixtured and spraying applied corrosion inhibitors was investigated. Moreover, when the corrosion action was under progress, the addition of spaying amino alcohol inhibitor was tested. Furthermore, attention was given to the condition that the specimens were exposed, partially immersed in a 3.5 wt% NaCl solution and in a saturated 3.5 wt% NaCl solution.

EXPERIMENTAL

Materials

To achieve the objectives of this investigation cylindrical reinforced cement mortar specimen were constructed. The test specimens were prepared with cement, sand and water in ratio 1:3:0.6. The cement type used was Greek Portland cement and its chemical analysis is shown in Table 1 and the water was tap from Athens water supply network appropriate for preparing specimens according to ELOT 452 [14]. Cylindrical steel rebars of type B500C with dimensions of 12 mm in diameter and 10 mm high were used for the reinforced test specimens. The rebars meet Greek specifications of Hellenic Organization for Standardization, ELOT 1421-3 [15].

Fabrication of the steel for the test specimens simply involved cutting to the consistent length of 100 mm. The test specimens considered for the present study were cylindrical 100 mm in height and 40 mm in diameter. Each contained one steel rebar in the position shown in Figure 1. The cement mortar constituents were mixed in a mortar mixer for approximately 5 minutes till a uniform consistency was achieved. The molds were filled with mortar and vibrated for consolidation using a vibrating table. Copper wire cables were connected to the steel bar for electrochemical measurements. Prior to the preparation, the steel surface was cleaned according to the ISO/DIS 8407.3 Standard [16]. In particular the surface of the steel bars was washed with water and then immersed in strong solution of HCl with organic corrosion inhibitor for 15 min, washed with water and finally washed thoroughly with distilled water to eliminate traces of the corrosion inhibitor and chloride ions. Following that, the surface was cleaned with alcohol and acetone and finally weighed to accuracy of 0.1 mg. Thereafter, the bars were placed in cylindrical moulds where the mortar was cast and stored at ambient conditions in the laboratory for 24 hours. After being demolded, the specimens were placed in water in curing room (RH > 98%, T = 20°C ± 1.5°C) for 24 hours and then kept for an additional 7 days at ambient temperature in a laboratory environment to stabilize internal humidity, followed by insulation with epoxy resin of the region shown in Figure 1. The experimental duration of this study was 200 days.

For the purpose of this work 4 series of specimens were partially immersed in a 3.5 wt% NaCl solution and 2 series remained fully immersed in a 3.5 wt% NaCl solution. The addition of the corrosion inhibitor was achieved according to the instruction given by the manufacturer. The analytical series tested in this work are presented in Table 2.

In Table 2 test series that were partially immersed in the solution are indicated with "A" whereas full immersion is characterized by using the letter "B" in the code name of the series.

Table 1: Chemical analysis of OPC

Oxide composition of cement (% by weight of cement)									
SiO_2	Al_2O_3	Fe_2O_3	CaO	MgO	K_2O	Na_2O	SO_3	LOI	CaO_f
20.67	4.99	3.18	63.60	2.73	0.37	0.29	2.414	2.52	2.41

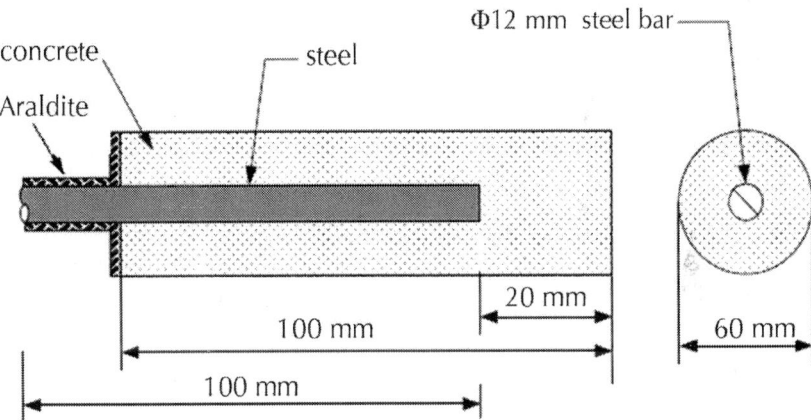

Figure 1: Schematic representation of a reinforced mortar specimen.

Table 2: Test series used in the present study

Code names of the series	Environmen	Characteristics
RSA	Partially immersed	Control specimens
DSA	Partially immersed	Inhibitor as admixture (4 wt% cement)
DSA14	Partially immersed	Inhibitor sprayed after 14 days from the day of manufacture
DSA3M	Partially immersed	Inhibitor sprayed after 3 months of exposure in the corrosive environment
RSB	Fully immersed	Control specimens
DSB	Fully immersed	Inhibitor as admixture (4 wt% cement)

Furthermore, "RS" stands for reference specimens and "DS" for data specimens where the inhibitor is either sprayed or as an admixture. Finally, the numbers on the code names show the time interval before the corrosion inhibitor was sprayed on the surface of the specimens.

The corrosion inhibitor used in this study was based on partially neutralized amino alcohols which is an adsorption inhibitor. Specifically, N-N-diethylaminoethanol is absorbed onto the steel surface forming organic films which protect both the anodic and cathodic sections of the rebar [17]. Besides, the transportation of surface applying amino alcohols inhibitor into concrete depends on the concrete quality, the porosity and the humidity [18, 19].

Methods

The following methods were employed in this study for the evaluation of the protective action of coatings against corrosion:

Half-cell Potential Measurements versus Time

During the exposure of the specimens in the corrosive environment the half-cell potential of steel rebars was periodically measured versus a saturated calomel electrode (SCE) and a high impedance voltmeter.

According to ASTM C876 Standard [20] corrosion potentials (E_{corr}) more negative than −350 mV, with respect to SCE, indicate greater than 90% probability of active reinforcement corrosion. Values less negative than −200 mV SCE indicate a probability of corrosion below 5%, while those falling between −200 and −350 mV SCE indicate uncertainty of corrosion.

Linear Polarization Technique (LPR)

The corrosion of reinforced concrete is an electrochemical process. For this reason it is possible to obtain the information on the state of corrosion with the application of external electrical stimulation and measurement of the subsequent electrical response. Hence, the linear polarization resistance method was used to measure corrosion rates over various time periods Although the corrosion of steel in concrete is an electrochemical process and does not obey to Ohm's law it has been shown that Ohms's law will be approximately true if polarization applied to the steel does not exceed ±20 mV. Thus the resulting current is linearly plotted versus potential. The test setup for the linear polarization resistance techniques (LPR) included a EG&G Model 263 Potensiostat/Galvanostat. Additionally the computer program Softcorr III developed by EG&G Princeton Research was used for applying the potential scan, analyzing the parameters of i_{corr} and R_p. The linear polarization resistance of steel reinforcement was evaluated at a scan rate of 0.15 mV/s. The corrosion current density was measured using the DC linear polarization resistance method. The resistance to polarization R_p ($\Omega \cdot cm^2$) was determined by conducting a linear polarization scan in the range of 20 mV of the open circuit potential and the corrosion current density i_{corr} [$\mu A/cm^2$] was then calculated using the Stern-Geary equation:

$$i_{corr} = \frac{B}{R_p \cdot A}$$

(1)

where B is a constant based on the anodic and cathodic Tafel constants where a value of 26 mV has been adopted for active corroding steel bars and 52 mV for passive conditions and A (cm^2) is the exposed area of the rebar.

This method has been considered to be a relatively simple and reliable technique to assess the rate of reinforcement corrosion in cement mortars. The corrosion level is considered negligible when i_{corr} is less than 0.1 µA/cm², it is considered low in the range between 0.1 and 0.5, moderate from 0.5 to 1 and high for values higher than 1 µA/cm² [21].

Weight Loss Measurements of Steel Rebars

The corrosion rate of reinforcing steel was determined by measuring the mass loss of the steel bars in different times. The steel bars were cleaned from any corrosion products according to the ISO/DIS 8407.3 Standard [16] and were weighted to accuracy of 0.1mg. The average mass loss was calculated from the difference between the initial and the final weight of each steel bar.

RESULTS AND DISCUSSION

Half-cell Potential Measurements

Half-cell potentials of the specimens that were partially immersed in 3.5 wt% NaCl solution for 200 days are shown in Figure 2. In the cement mortar specimens without corrosion inhibitor (RSA) the corrosion poten- tials were more negative than −300 mV SCE right from the initial stages of exposure and −620 mV SCE after 200 days of exposure.

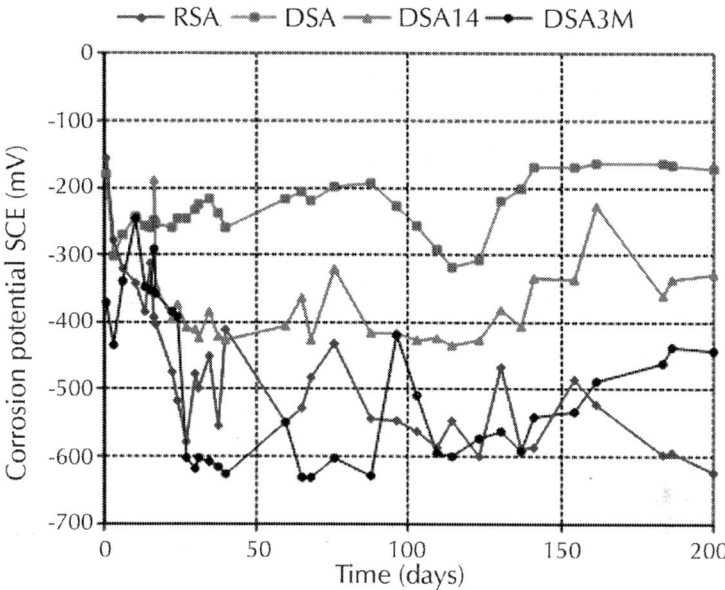

Figure 2: Half-cell potential measurements versus time in cement mortar specimens partially immersed in 3.5 wt% NaCl solution.

The half-cell potentials of specimens which an amino alcohol inhibitor was used as an admixture (DSA) show an increase of the potential values in the anodic direction during the first 90 days, attaining values around −195 mV. After that the potential slightly decays to values around −270 mV and finally reaches steady values around −170 mV indicating the passivity of reinforcing steel even after 200 days of exposure. The spraying applied amino alcohol corrosion inhibitor (DSA14) during the first 125 days of exposure remains practically constant with the time at values around −420 mV. However, after the first 125 days the half-cell potentials values are higher than the initial ones by −100 mV and remain in that value (−328 mV) until the end of the experiment. According to ASTM this is an indication of uncertainty of corrosion [20]. In the cement mortar specimens (DSA3M) that an amino alcohol inhibitor was applying by spraying at specimens' surface after 3 months the potential sharply decays to values around −620 mV. This decay of potential suggests loss of passivation. After the surface applying inhibitors at the 90th day there is a gradual increase in the half-cell potential values followed by stabilization at around −450 mV.

Despite the fact that these values are an indication of 90% probability of active reinforcement corrosion [20], the addition of the amino alcohol spaying inhibitor shows its effect on the cement mortar samples in the first days of the applying by significantly increase of the half-cell potential. The half-cell potentials results show that the effect of the inhibitor addition on the corrosion rate of reinforcing steel is dependent on the method of application. Thus, when the inhibitor is sprayed on specimens the corrosion inhibition is delayed comparatively to the situation in which the inhibitor is added into the cement mortar mixture.

Figure 3 depicts the half-cell potentials of control speci-mens and specimens that an amino alcohol inhibitor was used as an admixture. Both types were fully immersed in 3.5% wt. NaCl solution for 200 days. There is a tendency for the decreasing of potential from values range of −100: −200 mV to values of −600 mV for the both types of specimens.

The decrease in the specimens with amino alcohol corrosion inhibitor was more gradual compared to the rapid decrease of the specimens without corrosion inhibitor during the period 0 - 150 days. However, this is a very slightly indication of the better performance of the corrosion inhibitor because these values are below −350 mV. The amino alcohol corrosion inhibitor may displace chloride ion to create a passive film so the anodic reaction of iron are lower compared to the specimens without inhibitors during this period (0 - 150 days) [20]. Despite this dramatic decrease in the values of the half-cell potential for the both specimens probably the generation of ferrous oxide did not take place due to the absence of oxygen (fully immersed in 3.5% wt. NaCl solution). As a consequence, without oxygen caused by water saturation, iron will dissolve in excess of electron but will remain stable in solution as there is no compensating cathodic reaction so the potential is very negative. This shows the weakness of potential measurements in an environment without oxygen [22].

Linear Polarization Measurements

In previous works electrochemical experiments pointed out that amino alcohol corrosion inhibitor increased the corrosion resistance of steel samples immersed in solution simulating the concrete pore solution contaminated with chloride [23-30]. In fact, X-ray photoelectron spectroscopy analysis revealed that the inhibitors form a film containing nitrogen species, which seem to complex the chloride ions [23].

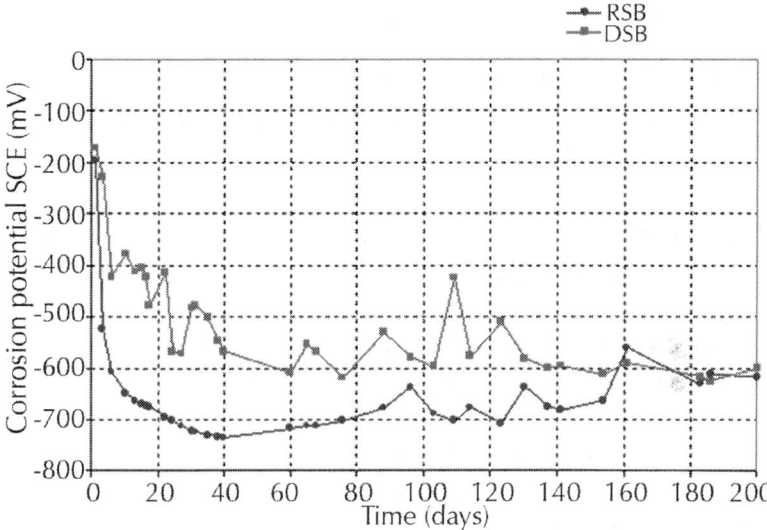

Figure 3: Half-cell potential measurements versus time in cement mortar specimens fully immersed in 3.5 wt% NaCl solution.

Previous different electrochemical measurements also showed that the amino alcohol inhibitors can reduce the corrosion rate even when steel is under corrosion attack [23-25]. Considering these results, the present work studies the ability of amino alcohol inhibitor to resist corrosion by measuring the corrosion current density in 3.5 wt% NaCl solution.

Figure 4 depicts the corrosion current density i_{corr} on rebars of the cement mortar specimens that were partially immersed in 3.5 wt% NaCl solution for 200 days. As it is shown i_{corr} values of the rebars without inhibitor increases with the exposure time from 2.2 µA/cm² to 4.5 µA/cm², values typical of high corrosion levels. It is also depicted that the usage of the corrosion inhibitor as an admixture (DSA) resulted in lower i_{corr} values (0.45 µA/cm²) which are indicative of low corrosion activity. For the specimens that the corrosion inhibitor was sprayed on the surface (DSA14), moderate corrosion values were achieved (0.78 µA/cm²).

These results show that the performance of the admixture amino alcohol inhibitor was better compared to the sprayed corrosion inhibitor. The admixture corrosion inhibitor occurs homogeneously inside the cement paste and as a result it is easier to generate the passive film to the surface of steel. Moreover, it is more effective in reducing the porosity of the cement paste by forming complex compounds with chlorides [8]. In contrast, the behavior of the sprayed corrosion inhibitor depends on the values of the diffusion coefficients of the inhibitor and the diffusion coefficient of chlorides.

The behavior of the sample that was kept 3 months in the corrosive environment with no inhibitor (DSA3M) in order to develop stronger corrosion activity and was afterwards sprayed with the inhibitor must be noted.

During the first period i_{corr} values slightly decrease from 3.2 µA/cm² to 3.1 µA/cm². After the inhibitor application in the surface of the specimen i_{corr} decreases to 0.86 µA/cm². This behaviour clearly shows that this inhibitor is able to decrease the corrosion rate of the steel even when it is initially under localized corrosion attack.

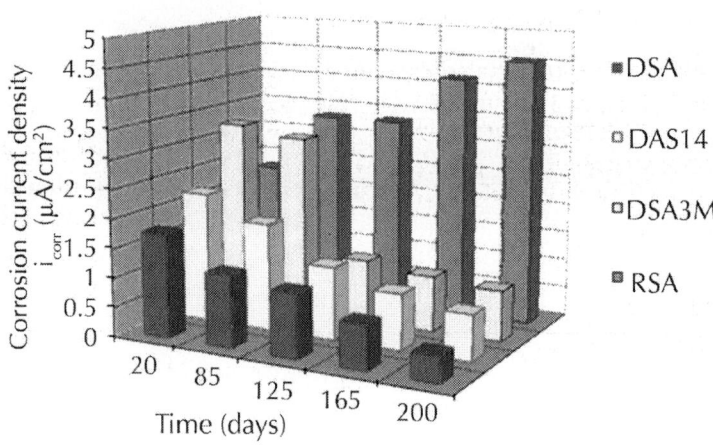

Figure 4: Corrosion current density versus time in cement mortar specimens partially immersed in 3.5 wt% NaCl solution.

This points out that the amino alcohol corrosion inhibitor is capable of diffusing through the cement and inhibiting the electrochemical corrosion process by passivation of the anodic and cathodic spots. This is an indication that a barrier layer is built up in the surface of the steel since i_{corr} values remain constant under the corrosive environment. Despite the significant reduction in the corrosion activity after the addition of the amino alcohol inhibitor, corrosion was not fully eliminated and a moderate level was achieved (0.86 µA/cm²).

Results of the linear polarization testing for control specimens and specimens with the corrosion inhibitor as an admixture that were fully immersed in 3.5% wt. NaCl solution for 200 days are presented in Figure 5.

Control specimens slightly decrease i_{corr} values from 0.87 µA/cm² to 0.82 µA/cm² after 200 days of exposure. When using the amino alcohol inhibitor as an admixture, i_{corr} values are reduced from 0.7782 µA/cm² to 0.6882 µA/cm². These values are lower than those obtained in the partially immersed specimens due to the saturated environment and to the absence of oxygen that restrains the corrosion process. This difference, after 200 days, is more significant for the control specimens. In fact, i_{corr} for the control specimens when are partially immersed is approximately 4.5 µA/cm² whereas in the case of full immersion is 0.8 µA/cm². However, the performance of the admixture inhibitor is slightly better under partially immersed solution compared to the performance under saturated environment.

Even under saline saturate environment the action of corrosion amino alcohol inhibitor is slightly better compared to the sample without inhibitor. This can be contributed to the formation of a complex compound (passive film) among the amino alcohol inhibitor, the chloride, the cement materials and the steel. As a result the anodic reaction of iron is in less extend. However, oxygen, a main parameter for the induced corrosion in the system is absent. Thus, the electrochemical corrosion process is not generated.

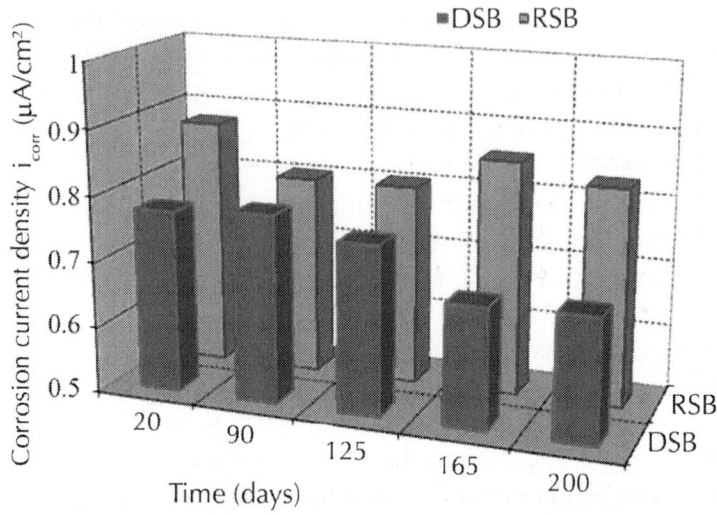

Figure 5: Corrosion current density versus time in cement mortar specimens fully immersed in 3.5 wt% NaCl solution.

Weight Loss Measurements of Corroded Steel Rebars

At the end of the exposure period the cement mortar specimens were carefully broken and the embedded steel bars were recovered. They were visually examined to assess their corrosion state qualitatively and then cleaned according to the ISO/DIS 8407.3 Standard [16] in an aqueous solution of hydrochloric acid containing a proprietary pickling restrainer which served to dissolve the corrosion products and cementitious debris without causing significant attack on the underlying steel.

Mass loss of all specimens that were partially immersed in the corrosive environment was obtained at 135, 165 and 200 days and the results are shown in Figure 6. Control specimens demonstrate higher mass loss values at all intervals which shows that in the absence of inhibitors the corrosion activity strongly increases with time, leading to higher mass loss. The specimens with addition of the corrosion inhibitor as admixture demonstrate a slightly increase of the mass loss versus time. The mean value of the mass loss at 200 days for the aforementioned specimens is 40 mg which is approximately similar to the value indicated by the specimens where the corrosion inhibitor was sprayed dafter 14 days.

The specimens that were sprayed with the corrosion inhibitor after 3 months of exposure attained a slow increase in the mass loss which is 44 mg after 200 days. These results show the inhibitory effects of amino alcohol corrosion inhibitor in reducing the corrosion even when it is under progress. In the case of the specimens that were fully immersed in 3.5 wt% NaCl solution both types of specimens showed similar trend in mass loss values and specifically after 200 days mas loss was 28.6 mg and 27.5 mg for control specimens and inhibitor amino alcohol admixture respectively as demonstrated in Figure 7.

It should be noted that the difference in the mass loss of the fully immersed specimens in 3.5 wt% NaCl solution both in control and in treated with corrosion inhibittor cement mortar specimens is not significantly important.

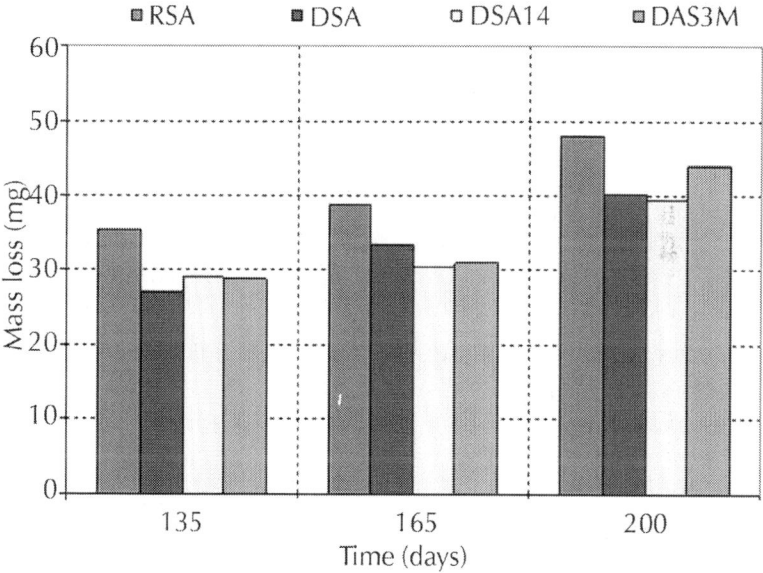

Figure 6: Mass loss of reinforcing steel bars time dependence partially immersed in 3.5 wt% NaCl solution.

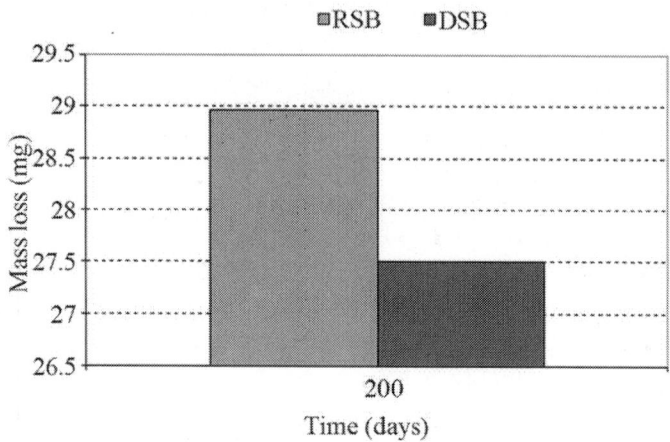

Figure 7: Mass loss of reinforcing steel bars time dependence fully immersed in 3.5 wt% NaCl solution.

This fact suggests that the absence of oxygen in the pores of cement mortar mass reduces the corrosion rates of steel reinforcements. This behaviour is in disagreement with the half-cell potential results which showed very negative results due to the absent of cathodic reaction. In fact, the specimens with the amino alcohol inhibitor show slightly more efficient in chloride attack than the control specimens. This difference can be explained as a result of inhibition of the anodic reaction on the steel due to the form of passive film by the amino alcohol. This could be an indication that the creation of the passive film could be independent on the presence of oxygen.

Moreover, the mass loss of the control specimens under partial immersed NaCl solution is dramatically higher compared with the mass loss at saturate NaCl solution. These results are in line with the results of the linear polarization analysis. In addition the mass loss of admixture inhibitor is also higher under partial immersed NaCl solution compared with the saturated environment. However, the results of the corrosion rate at linear polarization analysis showed slightly better performance of the admixture inhibitor at partially immersed NaCl solution.

The absent of oxygen in the saturated environment contribute to the elimination of the cathodic reaction so there is a possibility of a slightly wrong estimation of the corrosion current density.

CONCLUSIONS

The following conclusions can be drawn from this investigation:

- The amino alcohol types of corrosion inhibitors studied in this work (admixture and sprayed corrosion inhibitor) reduce the corrosion rate of the reinforcing steel when the specimens were partially immersed in a 3.5 wt% NaCl solution. In fact, according to the results of half-cell potential and linear polarization, admixture type is slightly more efficient in the cement mortar specimens than the surface applied amino alcohols corrosion inhibitors.

- In cement mortar specimens fully immersed in 3.5 wt% NaCl both control and admixture amino alcohol treated specimens demonstrate low corrosion activity because pores of cement mortar are filled with water and oxygen cannot penetrate.

- There is a decrease of the corrosion activity on steel in the addition of the amino alcohol spaying inhibitor in cement mortar samples even when the cement mortar specimens are initially under localized corrosion attack (partially immersed in a 3.5 wt% NaCl).

From the above, the goal of this paper to test the inhibitory action of the amino alcohol corrosion inhibitor under corrosive environments was achieved.

However, further research needs to be held regarding the mechanism that the amino alcohol corrosion inhibitor acts under the presence of high chloride concentration which diffuses in the cement. Moreover, the characteristic of the passive layer that is created by the inhibitor on the steel surface under different corrosive environment (partially corrosive environment or saturated environment) and under different addition method of the inhibitor (as admixture or sprayed) can be investigated.

REFERENCES

1. U. Nürberger, "Korrosionsschutz im Massivbau," Expert Verlag, Böbligen, 1991.

2. G. Batis and P. Pantazopoulou, "Advantages of the Simoultaneous Use Of Corrosion Inhibitors and Inorganic Coating," Cement and Concrete Technology in the 2000s Second International Symposium, Instabul, 6-10 September 2000, pp. 474-483.

3. M. J. Gaidis, "Chemistry of Corrosion Inhibitors," Cement & Concrete Composites, Vol. 26, No. 3, 2004, pp. 181-189. doi:10.1016/S0958-9465(03)00037-4

4. M. Pourbaix, "Atlas of Electrochemical Equilibria in Aqueous Solutions," Pergamon Press, Oxford, 1966.

5. J. P. Broomfield, "Assessing Corrosion Damage on Reinforced Concrete Structures," Proceedings of the International Conference Corrosion and Corrosion of Steel in Concrete, Sheffield, 24-28 July 1994, pp. 1-25.

6. C. M. Hansson, L. Mammoliti and B. B. Hope, "Corrosion Inhibitors in Concrete-Part I: The Principles," Cement and Concrete Research, Vol. 28, No. 12, 1998, pp. 1775-1781. doi:10.1016/S0008-8846(98)00142-2

7. L. Fedrizzi, F. Azzolini and P. L. Bonora, "The Use of Migrating Corrosion Inhibitors to Repair Motorways Concrete Structures Contaminated by Chlorides," Cement and Concrete Research Vol. 35, No. 3, 2005, pp. 551-561.doi:10.1016/j.cemconres.2004.05.018

8. M. Ormellese, M. Berra, F. Bolzoni and T. Pastore, "Corrosion Inhibitors for Chlorides Induced Corrosion in Reinforced Concrete Structures," Cement and Concrete Research, Vol. 36, No. 3, 2006, pp. 536-547. doi:10.1016/j.cemconres.2005.11.007

9. M. Duprat, N. Bui and F. Dabosi, "Corrosion-NACE," Vol. 35, No. 9, 1979, p. 392.doi:10.5006/0010-9312-35.9.392

10. M. Duprat and F. Dabosi, "Corrosion-NACE," Vol. 37, No. 2, 1981, p. 89.doi:10.5006/1.3593851

11. U. Maeder, "A New Class of Corrosion Inhibitors for Reinforced Concrete," Proceedings in the 9th Asian-Pacific Corrosion Control Conference, "Corrosion Prevention for Industrial Safety and Environmental Control", Taiwan, 1995, pp. 825-830.

12. A. Welle, J. Liao, K. Kaiser, M. Grunze, U. Mader and N. Blank, "Interactions of N,N-Dimethylaminoethanol with Steel Surfaces in Alkaline and Chloride Containing Solutions," Applied Surface

Science, Vol. 119, No. 3-4, 1997, pp. 185-190. doi:10.1016/S0169-4332(97)00216-X

13. I. Martínez, C. Andrade, N. Rebolledo, L. Luo and G. De Schutter, "Corrosion-Inhibitor Efficiency Control: Comparison by Means of Different Portable Corrosion Rate Meters," Corrosion: The Journal of Science and Engineering, Vol. 66, No. 2, 2010, pp. 026001-026001-12. doi:10.5006/1.3319663

14. Hellenic Organization for Standardization ELOT 452, "Determination of Total Hg Content to Water with Atomic Absorption Spectroscopy," Athens, 1983.

15. Hellenic Organization for Standardization ELOT 1421-3, "Steel for the Reinforcement of Concrete—Weldable Reinforcing Steel—Part 3: Technical Class B500C," Athens, 2005.

16. ISO/DIS 8407.3, "Procedures for removal of corrosion products from corrosion test specimen," Genève, Switzerland, 1986.

17. A. Routoulas, P. Pantazopoulou and G. Batis, "Evaluation of Parameters Influencing Reinforcement Corrosion by Means of a Strain Gauge Technique," Anti-Corrosion Methods and Materials, Vol. 50, No. 4, 2003, pp. 271- 279. doi:10.1108/00035590310482505

18. F. Wombacher, U. Maeder and B. Marazzani, "Aminoalcohol based mixed corrosion inhibitors," Cement & Concrete Composites, Vol. 26, No. 3, 2004, pp. 209-216.doi:10.1016/S0958-9465(03)00040-4

19. A. S. Abdulrahman, I. Mohammad and S. H. Mohammad, "Corrosion Inhibitors for Steel Reinforcement in Concrete: A Review," Scientific Research and Essays, Vol. 6, No. 20, 2011, pp. 4152-4162.

20. ASTM C 876-91, "Standard Test Method for Half-Cell Potential of Reinforcing Steel in Concrete," Annual Book of ASTM Standards. ASTM International, West Conshohocken, 1991.

21. European Concerted Action COST 509, "Corrosion and Protection of Metals in Contact with Concrete: Part 2. Monitoring," Final Report, European Commission, Brussel, 1997, p. 73.

22. J. P. Broomfield, "Condition Evaluation, Corrosion of Steel in Concrete," E&FN SPON, London, 1997. doi:10.4324/9780203414606

23.   H. E. Jamil, M. F. Montemor, R. Boulif, A. Shriri and M. G. S. Ferreira, "An Electochemica and Analytical Approach to the Inhibition Mechanism of an Amino-Alcohol-Based Corrosion Inhibitor for Reinforced Concrete," Electrochemica Acta, Vol. 48, No. 23, 2003, pp. 3509- 3518. doi:10.1016/S0013-4686(03)00472-9

24.   E. Jamil, A. Shriri, R. Boulif, A. C. Bastos, M. F. Montemor and M. G. S. Ferreira, "Electrochemical Behaviour of Amino Alchohol Based Inhibitors Used to Control Corrosion of Reinforcing Steel," Electrochemica Acta, Vol. 49, No. 17-18, 2004, pp. 2753-2760. doi:10.1016/j.electacta.2004.01.041

25.   H. E. Jamil, A. Shriri, R. Boulif, M. F. Montemor and M. G. S. Ferreira, "Corrosion Behaviour of Reinforcing Steel Exposed to an Amino Alcohol Based Corrosion Inhibitor," Cement and Concrete Composites, Vol. 27, No. 6, 2005, pp. 671-678.

26.   M. Sánchez and C. Alonso, "Accelerated Transport of Corrosion Inhibitors as Complementary Methodology for Electrochemical Chlorides Extraction Method," 2nd International Conference on Concrete Repair, Rehabilitation and Retrofitting, Cape Town, 24-26 November 2008, pp. 289-290.

27.   J. Kubo, S. Sawada, C. L. Page and M. M. Page, "Electrochemical Inhibitor Injection for Control of Reinforcement Corrosion in Carbonated Concrete," Materials and Corrosion, Vol. 59, No. 2, 2008, pp. 107-114. doi:10.1002/maco.200804161

28.   H. Zheng, W. Li, F. Ma and Q. Kong, "The Effect of a Surface-Applied Corrosion Inhibitor on the Durability of Concrete," Construction and Building Materials, Vol. 37, 2012, pp. 36-40. doi:10.1016/j.conbuildmat.2012.07.007

29.   M. Ormellese, F. Bolzoni, L. Lazzari and P. Pedeferri, "Effect of Corrosion Inhibitors on the Initiation of Chloride-Induced Corrosion on Reinforced Concrete Structures," Materials and Corrosion, Vol. 59, No. 2, 2008, pp. 98-106. doi:10.1002/maco.200804155

30.   T. A. Nguyen and X. Shi, "A Mechanistic Study of Corrosion Inhibiting Admixtures," Anti-Corrosion Methods and Materials, Vol. 56, No. 1, 2009, pp. 3-12.doi:10.1108/00035590910923400

Greener Approach towards Corrosion Inhibition

Neha Patni, Shruti Agarwal, and Pallav Shah

Department of Chemical Engineering, Institute of Technology, Nirma University, S. G. Highway, Ahmedabad, Gujarat 382481, India

ABSTRACT

Corrosion control of metals is technically, economically, environmentally, and aesthetically important. The best option is to use inhibitors for protecting metals and alloys against corrosion. As organic corrosion inhibitors are toxic in nature, so green inhibitors which are biodegradable, without any heavy metals and other toxic compounds, are promoted. Also plant products are inexpensive, renewable, and readily available. Tannins, organic amino acids, alkaloids, and organic dyes of plant origin have good corrosion-inhibiting abilities. Plant extracts contain many organic compounds, having polar atoms such as O, P, S, and N. These are adsorbed on the metal surface by these polar atoms, and protective films are formed, and various adsorption isotherms are obeyed. Various types of green inhibitors and their effect on different metals are mentioned in the paper.

INTRODUCTION

Corrosion is the deterioration of materials by chemical interaction with their environment. The term corrosion is sometimes also applied to the degradation of plastics, concrete, and wood, but generally refers to metals. The most widely used metal is iron (usually as steel). Corrosion can cause disastrous damage to metal and alloy structures causing economic consequences in terms of repair, replacement, product losses, safety, and environmental pollution. Due to these harmful effects, corrosion is an undesirable phenomenon that ought to be prevented [1]. There are several ways of preventing corrosion and the rates at which it can propagate with a view of improving the lifetime of metallic and alloy materials. The use of inhibitors for the control of corrosion of metals and alloys which are in contact with aggressive environment is one among the acceptable practices used to reduce and/or prevent corrosion. A corrosion inhibitor is a substance which, when added in small concentration to an environment, effectively reduces the corrosion rate of a metal exposed to that environment.

Corrosion inhibitors can be divided into two broad categories, namely, those that enhance the formation of a protective oxide film through an oxidizing effect and those that inhibit corrosion by selectively adsorbing on the metal surface and creating a barrier that prevents access of corrosive agents to the metal surface [1]. Almost all organic molecules containing heteroatoms such as nitrogen, sulphur, phosphorous, and oxygen show significant inhibition efficiency. Despite these promising findings about possible corrosion inhibitors, most of these substances are not only expensive but also toxic nonbiodegradable thus causing pollution problems. Hence, these deficiencies have prompted the search for their replacement.

Plants are sources of naturally occurring compounds, some with complex molecular structures and having different chemical, biological, and physical properties. The naturally occurring compounds are mostly used because they are environmentally acceptable, cost effective, and have abundant availability. These advantages are the reason for use of extracts of plants and their products as corrosion inhibitors for metals and alloys under different environment.

Different plant extracts can be used as corrosion inhibitors commonly known as green corrosion inhibitors. Some of them are the following. Tannins and their derivatives can be used to protect steel, iron, and other tools from corrosion. To protect mild steel in 2 M HCl solutions from corrosion, extracts from leaves can be used. Extracts of tobacco from twigs, stems, and leaves can protect steel and aluminium in saline solutions and strong pickling acids [1, 2]. Extracts from leaves were investigated and found to be effective corrosion inhibitors for mild steel in 2 M HCl solutions. Results for the same are shown in Figure 1.

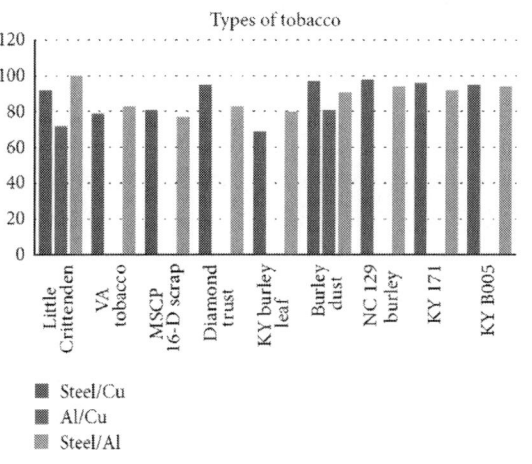

Figure 1: Inhibition efficiency of tobacco extracts for steel/Cu, Al/Cu, and steel/Al galvanic couples in 3.5% NaCl solution as measured by a zero resistance ammeter (ZRA) [1, 2].

It was found that maximum inhibition efficiency is 96% with only 0.01% tobacco concentration (100 ppm). Tobacco extracts contain high concentrations of chemical compounds such as alcohols, polyphenols, nitrogen-containing compounds, terpenes, carboxylic acids, and alkaloids that may exhibit electrochemical activity such as corrosion inhibition [1, 2].

Black pepper, Acacia gum, castor seed, and lignin are also good corrosion inhibitors for steel in acidic media [1, 3]. Mango peel extract is the most effective corrosion inhibitor for Al and Zn, and pomegranate fruit shells extract is most suitable for Cu. It was found that all extracts were more efficiently corrosion inhibitors in HCl solution as compared to H_2SO_4 solution [1, 4]. Aqueous extracts of

Eucalyptus leaves protect mild steel and copper in 1 M HCl solution from corrosion [1, 5]. Inhibition efficiency of plant extracts can be tested by various methods such as galvanostatic polarization, mass loss measurements, and surface characterization techniques. SEM studies provide the confirmatory evidence for the protection of mild steel by the inhibitor.

It was found that inhibition efficiency increases with increase in concentration of extract and decreases with increase in temperature.

Extract of leaves of Henna (Lawsonia) acts as a good corrosion inhibitor for carbon steel, nickel, and zinc in acidic, neutral, and alkaline solutions [1, 6]. The degree of inhibition depends on nature of metal and type of medium. For steel and nickel, the inhibition efficiency increased in the order: alkaline < neutral < acidic, while in case of zinc, it increased in the order: acid < alkaline < neutral, thereby reconciling with the observed concept of the Lawsonia extract being a mixed inhibitor [1, 6].

One among the crucial factors for the determination of the inhibition mechanism as well as the performance of the corrosion inhibitor is the solution pH. Most of the inhibitors are pH selective which depends on the molecular structure of the inhibitor, the metal corroding, the active species present in the solution, and the composition of the inhibitor.

Extract of Hibiscus sabdariffa can be used as corrosion inhibitor for mild steel in 2 M HCl and 1 M H_2SO_4 solution [1, 7]. Temperature changes do not affect inhibition performance of Hibiscus sabdariffa in 1 M H_2SO_4 solution.

The application of the acid extract of leaves of Citrus aurantifolia plant on the corrosion inhibition of mild steel in 1 M HCl solution was investigated using weight loss measurement and electrochemical studies [1, 8]. Inhibitive action of the same was tested on adsorption isotherms, and it was found to fit all the models tested, that is, Langmuir, Temkin, Freundlich, Frukin, and Flory - Huggins. This extract also acts as mixed-type inhibitor. A list of various plant materials that have been used as corrosion inhibitors is given in Table 1.

Table 1: Plant materials used as corrosion inhibitors [9]

S. no.	Metal	Medium	Inhibitor	Additive	Method	Findings	Reference
1	Zinc	2 M HCl	Aloe vera	—	Langmuir adsorption isotherm	A first-order kinetics relationship	[10]
2	Mild steel	H_2SO_4	Aloe vera		Infrared spectrophotometer, thermodynamic adsorption theories and gasometric (hydrogen evolution) methods. The study was conducted at 303 and 333 K	Chemical adsorption isotherm	[11]
3	Concrete steel surface	10 or 23 per cent sodium hydroxide	Banana plant juice taken from paradica and maghraby banana pseudostem		Weight loss method	Anticorrosive materials	[12]
4	Concrete steel surface		Magrabe banana stem		Galvanostatic polarization technique	Mechanical and physic-chemical properties	[13]
5	Mild steel	1 M HCl	Pennyroyal mint		Weight loss measurements, electrochemical polarization, and EIS methods	Cathodic inhibitor, adsorption isotherm	[14]
6	Mild steel	1 M HCl	Justicia gendarussaextract (JGPE)		Weight loss electrochemical techniques. AFM and ESCA	Mixed-type inhibitor. Obeys the Langmuir adsorption isotherm	[15]

7	Mild steel	0.1 M H_2SO_4	Caffeic acid	Weight loss, potentiodynamic polarisation, electrochemical impedance, and Raman spectroscopy	Controls the anodic reaction	[16]
8	Mild steel	1 M HCl and H_2SO_4	Combination of leaves and seeds (LVSD) extracts ofphyllanthus amarus	Weight loss and gasometric techniques	Temkin isotherm	[17]
9	Carbon steel	1 M HCl	Aqueous extracts of mango, orange, passion fruit, and cashew peels	Electrochemical impedance, spectroscopy, potentiodynamic polarization curves, weight loss measurements, and surface analysis	Langmuir adsorption isotherm	[18]
10	Carbon stee1	Ethanol	Caffeine	Voltammograms, Tafel plots, and EIS	The standard free energy of adsorption confirms a spontaneous chemical adsorption isotherm step	[19]
11	Al	0.5 M NaOH	Hibiscus sabdariffa leaves	Electrochemical measurements	Mixed-type inhibitor Langmuir and Dubinin Radushkevich isotherm	[20]
12	Al-Zn-Mg alloy	0.5 M NaOH	HibiscusTeterifa	Weight loss measurements	The adsorbed molecules of the alloy, lowers the corrosion rate.	[21]

13	Mild steel	H_2SO_4	Thyme, coriander, hibiscus, anise, black cumin, and garden cress	a.c., d.c., electrochemical techniques, and potentiodynamic polarization	Mixed-type inhibitor	[22]
14	Mild steel		Eucalyptus, hibiscus, and agaricus	Weight loss and polarization methods	Langmuir, Freundlich adsorption isotherm. Agaricus extract was found to be a cathodic inhibitor while extracts of eucalyptus and hibiscus were found to be mixed inhibitors	[23]
15	Mild steel	1 M HCl and 0.5 M H_2SO_4	Murraya Koenigii leaves	Weight loss, EIS, linear polarization, and potentiodynamic polarization techniques	Langmuir adsorption isotherm (Q, ΔH^*, and ΔS^*)	[24]
16	Mild steel	1 N HCl	Murraya Koenigii	Weight loss, gasometric studies, electrochemical polarization, AC impedance measurements, and SEM studies (30–80°C)	The protective film formed on the surface	[25]
17	Al	2 M HCl	Chromolaena odorata L.	Gasometric and thermometric techniques (30–60°C)	Langmuir adsorption isotherm	[26]

18	Mild steel	H_2SO_4	Ethanol extract of ITH-einsia crinata/IT		Weight loss, thermometric, hydrogen evolution techniques, and IR spectroscopy	Adsorption inhibitor Temkin and Frumkin adsorption	[27]
19	Mild steel	1 M HCl 0.5 H_2SO_4	Dacryodis edulis(DE)		Gravimetric and electrochemical techniques	DE extract was found to inhibit the uniform and localised corrosion of carbon steel in the acidic media	[28]
20	Al	HCl			Weight loss and hydrogen evolution methods	Langmuir adsorption isotherm, activation energies (E_a), activation enthalpy, and activation entropy	[29]
21	Al	0.5 M HCl	Azadirachta indica(AZI) plant	Iodide ions	Potentiodynamic polarization and impedance techniques	Freundlich adsorption isotherm	[30]
22	Mild steel	(60 ppm of Cl^-)	Aqueous extract of rhizome (Curcuma longaL.) powder	Zn^{2+}	Weight loss method, FTIR, UV fluorescence, and Electrochemical studies	Forms synergistic effect, protective film consists of a Fe^{2+}-curcumin complex and zinc hydroxide ($Zn[OH]_2$)	[31]
23	Al	HCl	Peepal (Ficus religiosa).		Mass loss and thermometric methods	IE dependent upon the concentrations of the inhibitor and the acid	[32]
25	Mild steel	0.1 M HCl	TL and BR inhibitors from green tea and rice bran		Weight loss method, polarization techniques	Cathodic inhibitor	[33]

26	Mild steel	0.2 M HCl	Bark and leaf solution extracts of mango (Mangifera indica)	Ambient temperature	Weight loss method	At 1.0 mL/100 mL of 0.2 M dilute sulphuric acid concentration gives good IE	[34]
27	Mild steel	HCl	Acid extract ofAndrographis paniculata		Mass loss method, Tafel polarization method, and impedance studies	Plant extract has the potential to serve as corrosion inhibitor	[35]
28	Al NaOH	Abrus precatorius	Ambient temperature		Weight loss and polarization techniques	Suitable adsorption isotherms were tested graphically	[36]
29	Mild steel	H_2SO_4	Combretum bracteosum		The gravimetric and hydrogen evolution measurements. Temp 30–60°C	Frumkin adsorption isotherm Kinetic parameters calculated, used in chemical cleaning and pickling	[37]
30	Al	1 M HCl	Root of ginseng		Weight loss techniques. Temp 30–60°C	IE 93.1% at 30°C at 50% v/v concentration of ginseng Freundlich adsorption isotherm, thermodynamic parameters calculated	[38]
31	Al	0.5 M NaOH and H_2SO_4	Vigna unguiculata(VU) extract		Weight loss techniques electrochemical studies. Temp 30 and 60°C	Freundlich and Temkin adsorption isotherms	[39]

32	Mild steel	1 M HCl	Mango, orange, passion fruit, and cashew peels	Electrochemical impedance spectroscopy, potentiodynamic polarization curves, weight loss measurements, and surface analysis	Langmuir adsorption isotherm, IE increases with increasing extract concentration and decreases with temperature	[40]
33	Mild steel	2 M HCl	olive (Olea europaea L.) leaves	Weight loss measurements, Tafel polarization, and cyclic voltammetry	Langmuir adsorption isotherm, olive extract decreases the charge density in the transpassive region	[41]
34	Mild steel	5% HCl	Both aqueous and alcoholic extracts of seven aloe plants	Weight loss measurements	IE 70–82%	[42]

Tannins, organic amino acids, alkaloids, and organic dyes of plant origin have good corrosion-inhibiting abilities. Plant extracts contain many organic compounds, having polar atoms such as O, P, S, and N. These are adsorbed on the metal surface by these polar atoms, and protective films are formed and various adsorption isotherms are obeyed.

The paper incorporates various types of green corrosion inhibitors and their effect on metals. Some important inhibitors in HCl solution, H_2SO_4 solutions, and water solution and effect of temperature and concentration of inhibitors on the process are discussed.

HCL SOLUTION AS MEDIUM

Grape Pomace for Carbon Steel

Acid solutions are widely used in industry, and some of the most important fields of application are acid pickling, chemical cleaning and processing, ore production, and oil well acidification [43–45]. C-steel is one of the most important alloys being used in a wide range of industrial applications. Corrosion problems arise as a result of the interaction between the aqueous solutions and C-steel, especially during the pickling process in which the alloy is brought in contact with highly concentrated acids. This process can lead to economic losses due to the corrosion of the alloy [43, 46]. The use of green inhibitors is one of the most practical ways possible for protecting carbon steel from corrosion.

Grape pomace is an industrial waste from wine and juice processing, and it primarily consists of grape seeds, skin and stems (~18–20 kg/100 kg of grapes) [43, 47–49]. It was found that grape pomace can effectively protect carbon steel from corrosion in 1 M HCl solution [43].

The inhibition efficiency of C-steel in 1 mol L^{-1} HCl increased with the concentration of crude and concentrated grape pomace extracts and was inversely associated with temperature. Presumably, the inhibitory effect was performed via the adsorption of compounds present in the grape pomace extracts onto the steel surface. Flavonoids are good candidates to explain the corrosion inhibition effects observed for grape pomace extracts. The adsorption of the grape pomace extracts followed a Langmuir adsorption isotherm. The Ea of C-steel dissolution increased in presence of the grape pomace extracts.

SEM revealed the persistence of a smooth surface on C-steel when grape pomace extracts were added, possibly due to the formation of an adsorptive film of phenolic compounds with electrostatic character [43].

Tannin for Mild Steel

Rhizophora racemosa is in abundance in the Mangrove forests of southern Nigeria. The bark of its stem is rich in tannins which can be described as any group of naturally occurring phenolic compounds. Their basic structure consists of garlic acid residues which are linked to glucose via glycosidic bonds [50, 51]. Thus tannins have an array of hydroxyl and carboxyl groups through which the molecules can adsorb on corroding metallic surfaces.

Ferrous materials, especially mild steel, on the other hand are largely used in acidic media in most industries including oil/gas exploration and ancillary activities. During such activities, inhibited hydrochloric acid is widely used in pickling, descaling, and stimulation of oil wells in order to increase oil and gas flow [50]. Tannins from Rhizophora Racemosa was found to be the most effective corrosion inhibitor for mild steel.

Studies on the corrosion behaviour of mild steel electrodes in inhibited hydrochloric acid are described. Conventional weight loss measurements show that a maximum concentration of 140 ppm of tannin from Rhizophora racemosa is required to achieve 72% corrosion inhibition. Similar concentration of tannin : H_3PO_4 in ratio 1 : 1 gave 61% inhibition efficiency, whereas efficiency obtained for phosphoric acid as inhibitor in the same environment was 55%. Corrosion rates obtained over six hours of exposure in 1 M HCl solution at inhibitor concentrations of 140 ppm are 2 mA/cm², 2.4 mA/cm², 2.6 mA/cm², and 6 mA/cm² for tannin, tannin/H_3PO_4, and H_3PO_4-inhibited and -uninhibited specimens respectively. Natural atmospheric exposure studies revealed that specimens treated in H_3PO_4 resisted corrosion for three weeks, while tannin-treated specimens suffered corrosion attack after one week of exposure tests [50].

Polyalthia longifolia for Mild Steel

Mild steel finds a lot of application in industries like metal finishing, boiler scale removal, pickling baths, and so forth. It gets rusted when it comes in contact with any acid. Acid solution, mostly HCl, is used to remove any undesirable scale or rust. Corrosion inhibitors are used to prevent the effect of corrosion in such cases. Use of hazardous chemical inhibitors is totally reduced because of environmental regulations. Chromates, phosphates, molybdates, and so forth and a variety of organic compounds containing heteroatoms like nitrogen, sulphur, and oxygen have been investigated as corrosion inhibitors [52–58].

The study shows that acid extract of Polyalthia longifolia (PL) is a good inhibitor for the corrosion of mild steel in HCl. The inhibition efficiency increases with the increase in inhibitor concentration and thus increases the protective action of the inhibitor on mild steel. The compound seems to function as inhibitor by being adsorbed on the metal surface. The inhibitor showed maximum inhibition efficiency of 87.79% at 1.5% v/v inhibitor concentration for an immersion period of 12 hours at 303 K. The % inhibition efficiency increases with increase in temperature, which confirms that PL acts as an effective inhibitor at high temperature also. The adsorption of acid extract of (PL) on the surface of mild steel is spontaneous, endothermic, and consistent with the isotherm models of Langmuir, Temkin, and Freundlich [52].

Flavin Mononucleotide (FMN) for Hot Rolled Steel

Heterocyclic compounds display potential properties for use as corrosion inhibitors due to the presence of nitrogen, oxygen, and sulphur in their ring structure [59–62]. In addition, planarity due to the presence of electrons and lone pairs of electrons on the heteroatoms contribute to their efficiency as inhibitors.

Flavin mononucleotide (7, 8-dimethyl-10-ribityl-isoalloxazine-5′ phosphate monosodium salt dihydrate) is a phosphate monosodium dihydrated salt of Vitamin B2 (Riboflavin). It consists of a heterocyclic isoalloxazine ring attached to the sugar alcohol, ribitol, which is derived from a D($-$) pentose sugar (ribose) that contains three antisymmetric carbons and a phosphate monosodium salt [59].

It was found that FMN is a potential inhibitor for corrosion of hot rolled steel in acidic medium. The inhibition efficiency of FMN increases with both concentration and temperature. The inhibitor follows Frumkin isotherm with negative values of ΔG°_{ads}, which signifies that the adsorption is a spontaneous process. High ΔG°_{ads} values indicate that the adsorption takes place by chemisorption at all temperatures except at the lowest temperature, where comprehensive adsorption exists. The Ea values for various concentrations of FMN are lower than Ea for acid, further confirming the role of chemisorption in the adsorption process. Quantum chemical analysis suggests that adsorption of FMN is mainly concentrated around the isoalloxazine ring [59].

WATER SOLUTIONS AS MEDIUM

Gum Exudates from Acacia Species (A. drepanolobium and A. senegal) for Mild Steel

Corrosion is a major destructive process affecting the performance of metallic materials in applications in many construction sectors. Corrosion is a naturally occurring phenomenon commonly defined as deterioration of metal surfaces caused by the reaction with the surrounding environmental conditions [63]. The use of the gum exudate from Acacia seyal var seyal as corrosion inhibitor for mild steel in fresh water has been reported [63, 64].

The study shows that gum exudates, from Acacia drepanolobium and Acacia senegal trees, which are natural products, inhibit the corrosion of mild steel in fresh water with A. senegal gum exhibiting better inhibition characteristics compared to Acacia drepanolobium. It was found that the inhibition performances of theAcacia gum exudates are insignificantly affected by temperature rise. Potentiodynamic polarization studies reveal that the gum exudates are mixed-type inhibitors of mild steel corrosion in fresh water with significant reduction of anodic current densities [63].

Asafoetida Extract (ASF) for Mild Steel in Sea Water

Asafoetida is an ingredient of a plant mixture reported to have antidiabetic properties in rats [65, 66]. Asafoetida has a broad range of uses in traditional medicine as an antimicrobial, antiepileptic, used for treating chronic bronchitis and whooping cough [65, 67, 68].

It was found that the formulation consisting of 4 mL of ASF and 25 ppm of Zn^{2+} offers 98% inhibition efficiency to carbon steel immersed in sea water. When immersion period increases, corrosion rate also increases. Polarization study reveals that this system formulation acts as a mixed type of inhibitor. The FTIR spectra reveal that the protecting film consists of Fe^{2+} Asafoetida (active ingredient) complex. AFM studies confirm that the surface is smoother. The smoothness of the surface is due to the formation of a compact protective film of Fe^{2+} ASF complex on the metal surface thereby inhibiting the corrosion of carbon steel [65].

Ginger Extract for Steel in Sulfide-Polluted Salt Water

Low-grade gram flour, natural honey, onion, potato, gelatin, plant roots, leaves, seeds, and flower gums are some of the good inhibitors. However, most of them have been tested on steel and nickel sheets. Although some studies have been performed on aluminum sheets, the corrosion effect is seen in very mild acidic or basic solutions (mill molar solutions) [69]. It was found that ginger can be effectively used to prevent corrosion of steel in sulphide-polluted salt water. Biological effect of ginger on Escherichia coli was also tested.

Ginger is suggested that it has oxygen donor atoms attached with the proteins and lipids on the bacterial tissues surface making a little activity for it. So it was observed that this inhibitor has no toxicity on the bacterial activity and can be applied on the waste water plants safely without any problems in treating waste water operations [69].

It was found that this extract inhibits the acid-induced corrosion of steel by virtue of adsorption of its components onto the metal surface. The inhibition process is a function of temperature, inhibitor concentration, and the metal as well as inhibitor adsorption abilities which is so much dependent on the number of adsorption sites. The mode of adsorption depends on the type of adsorption (physisorption and chemisorption) observed and could be attributed to the fact that this extract contains many different chemical compounds some of which can adsorb chemically and others adsorb physically. It may be due to the fact that adsorbed organic molecules can influence the behaviour of electrochemical reactions involved in corrosion processes in several ways [69].

Thus it was found that ginger acts as an inhibitor for corrosion of steel in sulfide-polluted salt water. The inhibition efficiency increases with increase in the concentration of the inhibitor. The inhibition is due to the adsorption of the inhibitor molecule on the metal surface by charge transfer or by the diffusion of the inhibitor molecules. The adsorption of these compounds on the metal surface follows Temkin adsorption isotherm. This inhibitor has no biological effect on the activity of Escherichia coli, and can be applied safely on waste water treatment plants.

H_2SO_4 SOLUTION AS MEDIUM

Tannin Extract of Chamaerops humilis (LF-Ch) Plant for Mild Steel

Tafel polarization curves and electrochemical impedance spectroscopy (EIS) approve that LF-Ch extract is an effective corrosion inhibitor for mild steel in 0.5 M sulfuric acid solution +5% EtOH. The inhibition efficiency improved with the increase of LF-Ch extract concentration, whether LF-Ch extract was used alone or in combination with KI. The increase in inhibitor efficiency is generated by the addition of KI to LF-Chextract. The Tafel polarization curves indicate that both LF-Ch extract is mixed anodic-cathodic type inhibitors. The addition of 0.025% KI to the solution leads to reduction in the essential usage of LF-Chextract to achieve desirable inhibition efficiency. The values of the inhibition efficiency increased with the immersion time and leads to the formation of a protective film which grows with increasing exposure time [70]. An inhibitor is usually added in small amount in order to slow down the rate of corrosion through the mechanism of adsorption [70–72].

Tryptamine (TA) as a Green Corrosion Inhibitor in Deaerated Sulfuric Acid

Tryptamine (TA), a derivative of the tryptophan, is relatively cheap, nontoxic and easy to produce in purity greater than 99% [73]. TA, a cheap molecule with a very low environmental impact, was found effective in inhibiting ARMCO iron corrosion in deaerated 0.5 M sulphuric acid in the 25–55°C temperature range. Results obtained from potentiodynamic polarisation and electrochemical impedance spectroscopy indicated that TA in the more concentrated solution and at 55°C also chemisorbs. EIS long-time tests (72 h and more) demonstrated that only the 10^{-2} M TA solution attained the maximum protection efficiency both at 25 and 55°C: IP ranged from about 95% to 98% [73].

Essential Oil of Salvia aucheri mesatlantica for Steel

Essential oil of aerial parts of Salvia aucheri Boiss. var. mesatlantica was obtained by hydrodistillation and analyzed by GC and GC/MS. The oil was predominated by camphor (49.59%). The inhibitory effect of this essential oil was estimated on the corrosion of steel in 0.5 M H_2SO_4 using electrochemical polarization and weight loss measurements. The corrosion rate of steel is decreased in the presence of natural oil [74]. Chemical analysis shows that camphor can be the major component of S. aucheri mesatlantica oil. Salvia aucheri mesatlantica oil mainly acts as good inhibitor for the corrosion of steel in 0.5 M H_2SO_4. Inhibition efficiency increases with both the concentration of inhibitor and the temperature. The natural oil acts on steel surface as anodic inhibitor. Inhibition efficiency on steel may occur by action of camphor [74].

CONCLUSIONS

Corrosion control of metals is technically, economically, environmentally, and aesthetically important. Corrosion of metals is the major problem in industries. Considering environmental and ecological reasons, green inhibitors are found to be effective. As organic corrosion inhibitors are toxic in nature, so green inhibitors which are biodegradable, without any heavy metals and other toxic compounds, are promoted. Also plant products are inexpensive, renewable, and readily available. The paper discusses some of the important inhibitors in HCl, water, and H_2SO_4 medium and effect of temperature and concentration of inhibitors on the process. Tannins, organic amino acids, alkaloids, and organic dyes of plant origin have good corrosion-inhibiting abilities. Plant extracts contain many organic compounds, having polar atoms such as O, P, S, and N. These are adsorbed on the metal surface by these polar atoms, and protective films are formed, and various adsorption isotherms are obeyed. Corrosion inhibitors can be divided into two broad categories, namely, those that enhance the formation of a protective oxide film through an oxidizing effect and those that inhibit corrosion by selectively adsorbing on the metal surface and creating a barrier that prevents access of corrosive agents to the metal surface. Inhibition efficiency depends on temperature and concentration of inhibitor. Some of the inhibitors are mixed-type inhibitors.

ACKNOWLEDGMENTS

The authors thank the Chemical Engineering Department of Institute of Technology, Nirma University, to provide infrastructure and ample resources needed for the work done.

REFERENCES

1.   J. Buchweishaija, "Phytochemicals as green corrosion inhibitors in various corrosive media a review," Chemistry Department, College of Natural and Applied Sciences, University of Dares Salaam.

2. G. D. Davis, Anthony Von Fraunhofer J, Krebs LA and Dacres CM, 1558, The use of Tobacco extracts as corrosion inhibitors. CORROSION, 2001.

3. K. Srivastava and P. Srivastava, "Studies on plant materials as corrosion inhibitors," British Corrosion Journal, vol. 16, no. 4, pp. 221–223, 1981. ·

4. R. M. Saleh, A. A. Ismail, and A. A. El Hosary, "Corrosion inhibition by naturally occurring substances. The effect of aqueous extracts of some leaves and fruit peels on the corrosion of steel, aluminum, zinc and copper in acids," British Corrosion Journal, vol. 17, no. 3, pp. 131–135, 1982. ·

5. K. Pravinar, A. Hussein, G. Varkey, and G. Singh, "Inhibition effect of aqueous extracts of Eucalyptusleaves on the acid corrosion of mild steel and copper," Transaction of the SAEST, vol. 28, no. 1, pp. 8–12, 1993.

6. A. Y. El-Etre, M. Abdallah, and Z. E. El-Tantawy, "Corrosion inhibition of some metals usingLawsonia extract," Corrosion Science, vol. 47, no. 2, pp. 385–395, 2005. View at Publisher

7. E. E. Oguzie, "Corrosion inhibitive effect and adsorption behaviour of Hibiscus sabdariffa extract on mild steel in acidic media," Portugaliae Electrochimica Acta, vol. 26, no. 3, pp. 303–314, 2008. ·

8. R. Saratha, S. V. Priya, and P. Thilagavathy, "Investigation of Citrus aurantiifolia leaves extract as corrosion inhibitor for mild steel in 1 M HCL," E-Journal of Chemistry, vol. 6, no. 3, pp. 785–795, 2009. ·

9. M. Sangeetha, S. Rajendran, T. S. Muthumegala, and A. Krishnaveni, Green corrosion inhibitors-An Overview.

10. O. K. Abiola and A. O. James, "The effects of Aloe vera extract on corrosion and kinetics of corrosion process of zinc in HCl solution," Corrosion Science, vol. 52, no. 2, pp. 661–664, 2010. View at Publisher· ·

11. N. O. Eddy and S. A. Odoemelam, "Inhibition of corrosion of mild steel in acidic medium using ethanol extract of Aloe vera," Pigment and Resin Technology, vol. 38, no. 2, pp. 111–115, 2009. View at Publisher

12. M. El-Sayed, O. Y. Mansour, I. Z. Selim, and M. M. Ibrahim,

"Identification and utilization of banana plant juice and its pulping liquor as anti-corrosive materials," Journal of Scientific and Industrial Research, vol. 60, no. 9, pp. 738–747, 2001. ·

13. S. H. Tantawi and I. Z. Selim, "Improvement of concrete properties and reinforcing steel inhibition using a natural product admixture," Journal of Materials Science and Technology, vol. 12, no. 2, pp. 95–99, 1996. ·

14. A. Bouyanzer, B. Hammouti, and L. Majidi, "Pennyroyal oil from Mentha pulegium as corrosion inhibitor for steel in 1 M HCl," Materials Letters, vol. 60, no. 23, pp. 2840–2843, 2006. View at Publisher

15. A. K. Satapathy, G. Gunasekaran, S. C. Sahoo, K. Amit, and P. V. Rodrigues, "Corrosion inhibition byJusticia gendarussa plant extract in hydrochloric acid solution," Corrosion Science, vol. 51, no. 12, pp. 2848–2856, 2009. View at Publisher

16. F. S. de Souza and A. Spinelli, "Caffeic acid as a green corrosion inhibitor for mild steel," Corrosion Science, vol. 51, no. 3, pp. 642–649, 2009. View at Publisher

17. P. C. Okafor, M. E. Ikpi, I. E. Uwah, E. E. Ebenso, U. J. Ekpe, and S. A. Umoren, "Inhibitory action ofPhyllanthus amarus extracts on the corrosion of mild steel in acidic media," Corrosion Science, vol. 50, no. 8, pp. 2310–2317, 2008. View at Publisher

18. J. C. da Rocha, J. A. da Cunha Ponciano Gomes, and E. D›Elia, "Corrosion inhibition of carbon steel in hydrochloric acid solution by fruit peel aqueous extracts," Corrosion Science, vol. 52, no. 7, pp. 2341–2348, 2010. View at Publisher

19. L. G. da Trindade and R. S. Gonçalves, "Evidence of caffeine adsorption on a low-carbon steel surface in ethanol," Corrosion Science, vol. 51, no. 8, pp. 1578–1583, 2009. View at Publisher

20. E. A. Noor, "Potential of aqueous extract of Hibiscus sabdariffa leaves for inhibiting the corrosion of aluminum in alkaline solutions," Journal of Applied Electrochemistry, vol. 39, no. 9, pp. 1465–1475, 2009. View at Publisher

21. F. A. Ayeni, V. S. Aigbodion, and S. A. Yaro, "Non-toxic plant extract as corrosion inhibitor for chill cast Al-Zn-Mg alloy in caustic soda solution," Eurasian Chemico-Technological Journal, vol. 9, no. 2, pp. 91–96, 2007. ·

22. E. Khamis and N. Alandis, "Herbs as new type of green inhibitors for acidic corrosion of steel,"Materialwissenschaf Tund Werkstoffiechnik, vol. 33, no. 9, pp. 550–554, 2002.

23. A. Minhaj, P. A. Saini, M. A. Quraishi, and I. H. Farooqi, "A study of natural compounds as corrosion inhibitors for industrial cooling systems," Corrosion Prevention and Control, vol. 46, no. 2, pp. 32–38, 1999. ·

24. M. A. Quraishi, A. Singh, V. K. Singh, D. K. Yadav, and A. K. Singh, "Green approach to corrosion inhibition of mild steel in hydrochloric acid and sulphuric acid solutions by the extract of Murraya koenigii leaves," Materials Chemistry and Physics, vol. 122, no. 1, pp. 114–122, 2010. View at Publisher· ·

25. A. Sharmila, A. A. Prema, and P. A. Sahayaraj, "Influence of Murraya koenigii (curry leaves) extract on the corrosion inhibition of carbon steel in HCL solution," Rasayan Journal of Chemistry, vol. 3, no. 1, pp. 74–81, 2010. ·

26. I. B. Obot and N. O. Obi-Egbedi, "An interesting and efficient green corrosion inhibitor for aluminium from extracts of Chlomolaena odorata L. in acidic solution," Journal of Applied Electrochemistry, vol. 40, no. 11, pp. 1977–1984, 2010. View at Publisher

27. N. O. Eddy and A. O. Odiongenyi, "Corrosion inhibition and adsorption properties of ethanol extract of ITHeinsia crinata/IT on mild steel in H_2SO_4," Pigment and Resin Technology, vol. 39, no. 5, pp. 288–295, 2010. View at Publisher

28. E. E. Oguzie, C. K. Enenebeaku, C. O. Akalezi, S. C. Okoro, A. A. Ayuk, and E. N. Ejike, "Adsorption and corrosion-inhibiting effect of Dacryodis edulis extract on low-carbon-steel corrosion in acidic media," Journal of Colloid and Interface Science, vol. 349, no. 1, pp. 283–292, 2010. View at Publisher ··

29. E. I. Ating, S. A. Umoren, I. I. Udousoro, E. E. Ebenso, and A. P. Udoh, "Leaves extract of ananas sativum as green corrosion inhibitor for aluminium in hydrochloric acid solutions," Green Chemistry Letters and Reviews, vol. 3, no. 2, pp. 61–68, 2010. View at Publisher

30. S. T. Arab, A. M. Al-Turkustani, and R. H. Al-Dhahiri, "Synergistic effect of Azadirachta Indica extract and iodide ions on the corrosion inhibition of aluminium in acid media," Journal of the

Korean Chemical Society, vol. 52, no. 3, pp. 281–294, 2008. ·

31. S. Rajendran, S. Shanmugapriya, T. Rajalakshmi, and A. J. Amal Raj, "Corrosion inhibition by an aqueous extract of rhizome powder," Corrosion, vol. 61, no. 7, pp. 685–692, 2005. ·

32. T. Jain, R. Chowdhary, P. Arora, and S. P. Mathur, "Corrosion inhibition of aluminum in hydrochloric acid solutions by peepal (Ficus Religeosa) extracts," Bulletin of Electrochemistry, vol. 21, no. 1, pp. 23–27, 2005. ·

33. Z. Liu and G.-L. Xiong, "Preparation and application of plant inhibitors," Corrosion and Protection, vol. 24, no. 4, pp. 146–150, 2003. ·

34. C. A. Loto, "The effect of mango bark and leaf extract solution additives on the corrosion inhibition of mild steel in dilute sulphuric acid—part I," Corrosion Prevention and Control, vol. 48, no. 1, pp. 38–41, 2001. ·

35. S. P. Ramesh, K. P. Vinod Kumar, and M. G. Sethuraman, "Extract of andrographis paniculata as corrosion inhibitor of mild steel in acid medium," Bulletin of Electrochemistry, vol. 17, no. 3, pp. 141–144, 2001. ·

36. R. Rajalakshmi, S. Subhashini, M. Nanthini, and M. Srimathi, "Inhibiting effect of seed extract ofAbrus precatorius on corrosion of aluminium in sodium hydroxide," Oriental Journal of Chemistry, vol. 25, no. 2, pp. 313–318, 2009. ·

37. P. C. Okafor, I. E. Uwah, O. O. Ekerenam, and U. J. Ekpe, "Combretum bracteosum extracts as eco-friendly corrosion inhibitor for mild steel in acidic medium," Pigment and Resin Technology, vol. 38, no. 4, pp. 236–241, 2009. View at Publisher

38. I. B. Obot and N. O. Obi-Egbedi, "Ginseng root: a new efficient and effective eco-friendly corrosion inhibitor for aluminium alloy of type AA 1060 in hydrochloric acid solution," International Journal of Electrochemical Science, vol. 4, no. 9, pp. 1277–1288, 2009. ·

39. S. A. Umoren, I. B. Obot, L. E. Akpabio, and S. E. Etuk, "Adsorption and corrosive inhibitive properties of Vigna unguiculata in alkaline and acidic media," Pigment and Resin Technology, vol. 37, no. 2, pp. 98–105, 2008. View at Publisher

40. J. C. da Rocha, J. A. da Cunha Ponciano Gomes, and E. D›Elia, "Corrosion inhibition of carbon steel in hydrochloric acid solution by fruit peel aqueous extracts," Corrosion Science, vol. 52, no. 7, pp. 2341–2348, 2010. View at Publisher

41. A. Y. El-Etre, "Inhibition of acid corrosion of carbon steel using aqueous extract of olive leaves,"Journal of Colloid and Interface Science, vol. 314, no. 2, pp. 578–583, 2007. View at Publisher

42. R. M. Saleh, M. A. Abd El Alim, and A. A. El Hosary, "Corrosion inhibition by naturally occurring substances: constitution and inhibiting property of Aloe plants," Corrosion Prevention and Control, vol. 30, no. 1, pp. 9–10, 1983. ·

43. J. C. Da Rocha, G. J. A. Ponciano C, E. D. Elia et al., "Grape pomace extracts as green corrosion inhibitors for carbon steel in hydrochloric acid solutions," International Journal of Electrochemical Science, vol. 7, pp. 11941–11956.

44. X.-H. Li, S.-D. Deng, and H. Fu, "Inhibition by Jasminum nudiflorum Lindl. leaves extract of the corrosion of cold rolled steel in hydrochloric acid solution," Journal of Applied Electrochemistry, vol. 40, no. 9, pp. 1641–1649, 2010. View at Publisher ·

45. A. U. Ezeoke, O. G. Adeyemi, O. A. Akerele, and N. O. Obi-Egbedi, "Computational and experimental studies of 4-aminoantipyrine as corrosion inhibitor for mild steel in sulphuric acid solution,"International Journal of Electrochemical Science, vol. 7, no. 1, pp. 534–553, 2012. ·

46. A. Y. El-Etre, "Inhibition of C-steel corrosion in acidic solution using the aqueous extract of zallouh root," Materials Chemistry and Physics, vol. 108, no. 2-3, pp. 278–282, 2008. View at Publisher ·

47. L. M. A. S. de Campos, F. V. Leimann, R. C. Pedrosa, and S. R. S. Ferreira, "Free radical scavenging of grape pomace extracts from Cabernet sauvingnon (Vitis vinifera)," Bioresource Technology, vol. 99, no. 17, pp. 8413–8420, 2008. View at Publisher

48. M. Spanghero, A. Z. M. Salem, and P. H. Robinson, "Chemical composition, including secondary metabolites, and rumen fermentability of seeds and pulp of Californian (USA) and Italian grape pomaces," Animal Feed Science and Technology, vol. 152, no. 3-4, pp. 243–255, 2009. View at Publisher· ·

49. M. A. Bustamante, R. Moral, C. Paredes, A. Pérez-Espinosa, J. Moreno-Caselles, and M. D. Pérez-Murcia, Waste Management, 2008.

50. M. Oki, E. Charles, C. Alaka, and T. K. Oki, Corrosion Inhibition of Mild Steel in Hydrochloric Acid By Tannins From Rhizophora Racemosa Materials Sciences and Applications, vol. 2, 2011.

51. G. I. Nonaka, "The isolation and structure elucidation of tannins," Pure and Applied Chemistry, vol. 6, no. 3, pp. 357–360, 1989.

52. V. G. Vasudha and K. Shanmuga Priya, "Polyalthia longifolia as a corrosion inhibitor for mild steel in HCl solution," Research Journal of Chemical Sciences, vol. 3, no. 1, pp. 21–26, 2013.

53. S. A. M. Refaey, "Inhibition of steel pitting corrosion in HCl by some inorganic anions," Applied Surface Science, vol. 240, no. 1–4, pp. 396–404, 2005. View at Publisher

54. M. A. Quraishi and H. K. Sharma, "Thiazoles as corrosion inhibitors for mild steel in formic and acetic acid solutions," Journal of Applied Electrochemistry, vol. 35, no. 1, pp. 33–39, 2005. View at Publisher

55. H. Ashassi-Sorkhabi, B. Shaabani, and D. Seifzadeh, "Corrosion inhibition of mild steel by some schiff base compounds in hydrochloric acid," Applied Surface Science, vol. 239, no. 2, pp. 154–164, 2005. View at Publisher

56. M. Bouklah, A. Ouassini, B. Hammouti, and A. El Idrissi, "Corrosion inhibition of steel in sulphuric acid by pyrrolidine derivatives," Applied Surface Science, vol. 252, no. 6, pp. 2178–2185, 2006. View at Publisher

57. E. E. Oguzie, B. N. Okolue, E. E. Ebenso, G. N. Onuoha, and A. I. Onuchukwu, "Evaluation of the inhibitory effect of methylene blue dye on the corrosion of aluminium in hydrochloric acid," Materials Chemistry and Physics, vol. 87, no. 2-3, pp. 394–401, 2004. View at Publisher

58. S. A. Ali, M. T. Saeed, and S. U. Rahman, "The isoxazolidines: a new class of corrosion inhibitors of mild steel in acidic medium," Corrosion Science, vol. 45, no. 2, pp. 253–266, 2003. View at Publisher ··

59. S. M. Bhola, G. Singh, and B. Mishra, "Flavin mononucleotide

as a corrosion inhibitor for hot rolled steel in hydrochloric acid," International Journal of Electrochemical Science, vol. 8, pp. 5635–5642, 2013.

60. M. A. Quraishi and R. Sardar, "Dithiazolidines—a new class of heterocyclic inhibitors for prevention of mild steel corrosion in hydrochloric acid solution," Corrosion, vol. 58, no. 2, pp. 103–107, 2002.·

61. S. L. Granese, B. M. Rosales, C. Oviedo, and J. O. Zerbino, "The inhibition action of heterocyclic nitrogen organic compounds on Fe and steel in HCl media," Corrosion Science, vol. 33, no. 9, pp. 1439–1453, 1992. ·

62. S. N. Banerjee and S. Misra, "1,10,-phenanthroline as corrosion inhibitor for mild steel in sulfuric acid solution," Corrosion, vol. 45, no. 9, pp. 780–783, 1989. ·

63. J. Buchweishaija, Plants As a Source of Green Corrosion Inhibitors: The Case of Gum Exudates From Acacia Species, Chemistry Department, College of Natural and Applied Science.

64. J. Buchweishaija and G. S. Mhinzi, "Natural products as a source of environmentally friendly corrosion inhibitors: the case of gum exudate from Acacia seyal var. seyal," Portugaliae Electrochimica Acta, vol. 26, no. 3, pp. 257–265, 2008. ·

65. M. Sangeetha, S. Rajendran, J. Sathiyabama, and P. Prabhakar, "Asafoetida extract (ASF) as green corrosion inhibitor for mild steel in sea water," International Research Journal of Environment Sciences, vol. 1, no. 5, pp. 14–21, 2012.

66. F. M. Al-Awadi, M. A. Khattar, and K. A. Gumaa, "On the mechanism of the hypoglycaemic effect of a plant extract," Diabetologia, vol. 28, no. 7, pp. 432–434, 1985. ·

67. K. Srinivasan, "Role of spices beyond food flavoring: nutraceuticals with multiple health effects," Food Reviews International, vol. 21, no. 2, pp. 167–188, 2005. View at Publisher

68. M. Z. Abdin and Y. P. Abdin, Abrol, ISBN 81-7319-707-5, Published Alpha Science IntﺑL Ltd. TraditionaL Systems of Medicine, 2005.

69. A. E. -A. S. Fouda, A. A. Nazeer, M. Ibrahim, and M. Fakih, "Ginger extract as green corrosion inhibitor for steel in sulfide

polluted salt water," Journal of the Korean Chemical Society, vol. 57, no. 2, pp. 272–278, 2013.

70.   O. Benali, H. Benmehdi, O. Hasnaoui, C. Selles, and R. Salghi, "Green corrosion inhibitor: inhibitive action of tannin extract of Chamaerops humilis plant for the corrosion of mild steel in 0. 5M H_2SO_4,"Journal of Materials and Environmental Science, vol. 4, no. 1, pp. 127–138, 2013.

71.   N. O. Eddy, "Inhibitive and adsorption properties of ethanol extract of Colocasia esculenta leaves for the corrosion of mild steel in H_2SO_4," International Journal of Physical Sciences, vol. 4, no. 4, pp. 165–171, 2009.

72.   A. Bouyanzer and B. Hammouti, "A study of anti-corrosive effects of Artemisia oil on steel," Pigment and Resin Technology, vol. 33, no. 5, pp. 287–292, 2004. View at Publisher

73.   G. Moretti, F. Guidi, and G. Grion, "Tryptamine as a green iron corrosion inhibitor in 0.5 M deaerated sulphuric acid," Corrosion Science, vol. 46, no. 2, pp. 387–403, 2004. View at Publisher

74.   M. Znini, L. Majidi, A. Bouyanzer et al., "Essential oil of Salvia aucheri mesatlantica as a green inhibitor for the corrosion of steel in 0.5 M H_2SO_4," Arabian Journal of Chemistry, vol. 5, no. 4, pp. 467–474, 2010. View at Publisher

Study of Flow-assisted Corrosion of AZ91D Magnesium Alloy in Loop System Based on Array Electrode Technology

Hualiang Huang[1,2], Guoan Zhang[1], Jiakuan Yang[1], Zhiquan Pan[2], and Xingpeng Guo[1]

[1]Hubei Key Laboratory of Materials Chemistry and Service Failure, School of Chemistry and Chemical Engineering, Huazhong University of Science and Technology, Wuhan 430074, China

[2]School of Chemistry and Environmental Engineering, Wuhan Institute of Technology, Wuhan 430073, China

ABSTRACT

A loop system was used to investigate flow-assisted corrosion (FAC) of AZ91D magnesium alloy at an elbow based on array electrode technology by potentiodynamic polarization, computational fluid dynamics, simulation and surface analysis. The experimental results demonstrate the fluid hydrodynamics plays a significant role in the FAC of AZ91D magnesium alloy. The corrosion rate increases from the outer wall to the inner wall of the elbow, with the higher corrosion rate corresponding to the higher flow velocity and larger shear stress at the elbow. The maximum corrosion rate appears at the innermost wall of the elbow, the location with the maximum flow velocity and shear stress.

INTRODUCTION

Magnesium alloys appear to be promising alternatives to aluminum alloys and steel used in the automotive industry because of their light density, high strength-to-weight ratio, and good mechanical properties [1, 2], especially in the cooling system of an engine block. However, the corrosion of magnesium alloys is a serious problem in the cooling system of an engine block [3], which restricts their applications. During the past decade, there were many research works which focused on the corrosion behaviour and mechanism of magnesium alloys [4–11].

AZ91D magnesium alloy is one of the most popular magnesium alloys used today. Although many works have been done to investigate its corrosion behaviour and mechanism in a static medium [12], there is no report about its corrosion behaviour and mechanism in a flow medium. Since the fluid flow has significant effects on the mass transfer process and the removal of corrosion products on the electrode surface [13], the different corrosion behaviour of AZ91D magnesium alloy is expected in a flow medium.

Flow-assisted corrosion (FAC), caused by the combined action of corrosion and fluid flow, is one of the main reasons resulting in failure of heat exchanger [14]. At present, rotating disk electrode (RDE) or rotating cylinder electrode (RCE) [15–17], impingement jet systems [18–24], and loop systems [13, 25, 26] have been used extensively for FAC investigation. However, RDE, RCE, and impingement jet systems could not really reflect the flow pattern of fluid in a pipe, especially at an elbow. Therefore, a loop system should be applied to simulate the realistic flow environments in a pipe.

Elbow is an important part of heat exchanger configuration. However, the flow pattern will occur to great changes in flow direction and flow velocity in a 90° elbow, resulting in significant difference in the corrosion behaviour at different locations of the elbow [27]. Due to the sudden change in the flow pattern, the wall thinning is further exacerbated by FAC at the elbow. Therefore, FAC at the elbow is rather serious among the damage of heat exchanger. Apparently, there should be correlation between the corrosion behaviour at different locations of the elbow and the flow pattern. However, there have been few works to study the different corrosion behaviour at different locations of an elbow [13]. Array electrode technique, a configuration of multielectrode system, can be used for studying the different corrosion behaviour at different locations of the elbow.

Based on array electrode technology, a loop system was designed and used to investigate the FAC of AZ91D magnesium alloy at a 90° elbow in circulating ASTM D1384-87 solution by potentiodynamic polarization, computational fluid dynamics (CFD) simulation, and surface analysis in this work. The corrosion behaviour and mechanism at different locations of the elbow were investigated to determine the effect of fluid hydrodynamics and the correlation between the corrosion behaviour and the distributions of flow velocity and shear stress at a 90° elbow.

EXPERIMENTAL

Loop System for FAC Test

Figure 1(a) shows the circulating loop system used for FAC test. It consisted of pipes, a centrifugal pump, a container, a pressure gage, a flow meter, and array electrode test section. The solution was supplied from a 125L container and circulated through the centrifugal pump. The flow velocity was controlled by controlling the pump rotational speed using a controller. The flow velocity in this study is 5.31 m/s, which was measured by the flow meter. The loop system was made of 316L stainless steel pipe with the inner diameter of 50 mm. After pretreatment, array electrodes were mounted into the elbow test section with the same spacing distance in flow direction. Figure 1(b) shows the schematic diagram of the test section. Figure1(c) shows the photograph of the test section with 20 specimens at the elbow. The exposed surfaces of array electrodes were in accordance with the internal surface of pipe, as shown in Figure 1(d). The 20 specimens at the elbow are symmetrical with respect to the central plane of pipe. In addition, a specimen (electrode 1) is in the straight pipe before the outlet of the elbow, and one specimen (electrode 7) is in the straight pipe after the inlet of the elbow.

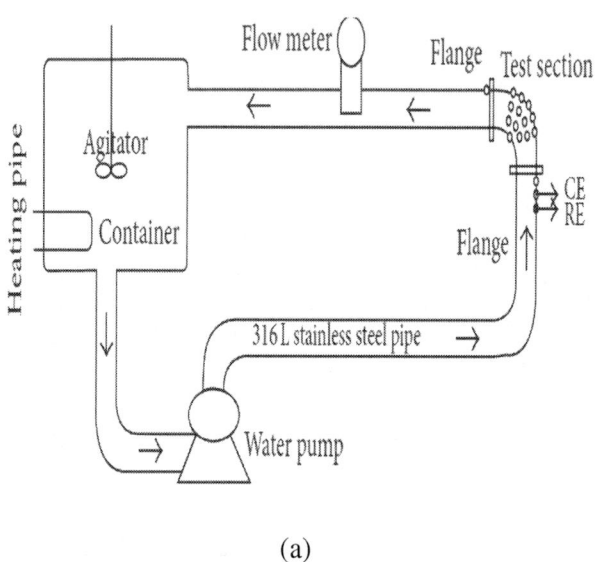

(a)

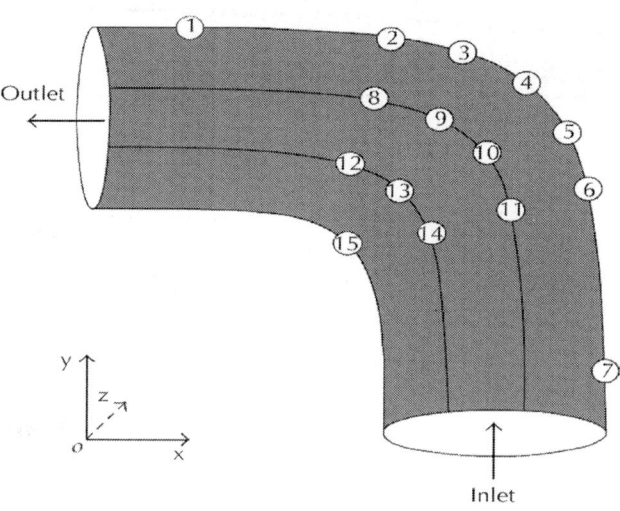

(b)

(c)

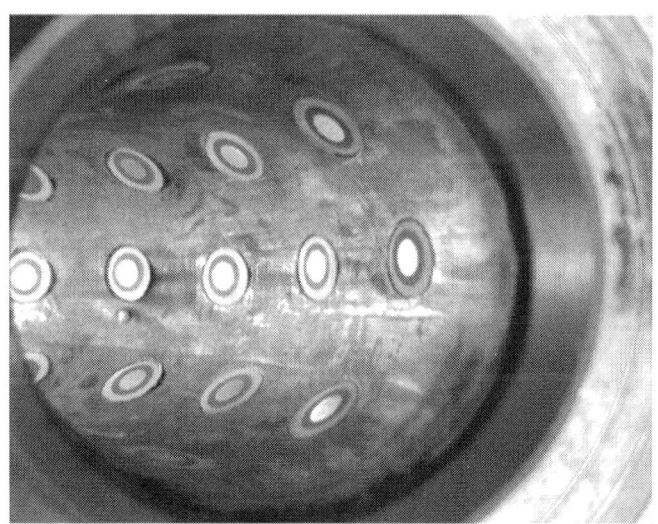

(d)

Figure 1: Schematic diagram of FAC loop test system and array electrodes: (a) loop test system, (b) distribution of array electrodes in test section, (c) the assembly of elbow test section, and (d) distribution of array electrodes at the inner wall of elbow.

Electrode and Solutions

The array electrodes were made up of AZ91D magnesium alloy with identical diameter of 5 mm (surface area of 19.625 mm^2) and height of 60 mm. The specimens were embedded in epoxy resin, leaving a working area of 19.625 mm^2. The working surface was polished with 1200 grit emery papers and then cleaned by distilled water and pure ethanol.

The test solution is ASTM D1384-87 solution containing 148 mg/L (1.04 mM) Na_2SO_4 + 138 mg/L (1.64 mM) $NaHCO_3$ + 165 mg/L (2.82 mM) NaCl, which was made up from analytical grade reagents and deionized water. The pH of the solution is 8.2.

Potentiodynamic Polarization Measurements

An electrochemical test system was used for potentiodynamic polarization curve measurement during the FAC test. A three-electrode electrochemical cell was constructed with array electrodes as working electrodes (WE), a platinum plate as counter electrode (CE), and a saturated calomel electrode (SCE) as reference electrode (RE), and CE and RE were located at the straight pipe before the inlet of the elbow and closer to the array electrodes, as shown in Figure 1(a). Potentiodynamic polarization curve was measured at a potential sweep rate of 1 mV/s after the working electrode had reached a steady state in the solution [28]. For the reproducibility, the polarization curve measurements were repeated more than three times. The test was performed with the flow velocity of 5.31 m/s at room temperature (about 25°C) and an atmospheric pressure.

Surface Morphology Analysis after FAC Test

After FAC test, the surface morphologies of array electrodes were observed by an optical microscope (VHX-1000E, Keyence, Japan).

CFD Simulation

Professional fluid simulation software Fluent was employed to perform CFD simulation. Preprocessing software Gambit was used to establish the geometric model. The straight section before the inlet of the elbow was set as 0.5 m and the straight section before the outlet of the elbow was also set as 0.5 m. Volume meshes were constructed with the interval size of 0.01 m. A flow velocity of 5.31 m/s at the inlet and an atmospheric pressure (101325 Pa) at the outlet were set as the boundary conditions. The fluid was assumed to be incompressible and a k-ε turbulent model (double equation model) was used to numerically solve the simulation since the fluid flowed at a Reynolds number of 265500 (calculated according to the inner diameter of pipe and flow velocity). The Reynolds number was much higher than 4000, indicating a turbulent flow. Turbulence intensity in the simulation was 3.36% (calculated according to the calculated Reynolds number). The k-ε turbulence equation was solved by iterative method with a convergence criterion of 0.00001.

RESULTS

Potentiodynamic Polarization Curve Measurement

Potentiodynamic polarization curves of AZ91D magnesium alloy array electrodes under the flow condition with 5.31 m/s are shown in Figure 2. It is seen from Figure 2 that these electrodes are in an active dissolution state under the flow condition, and the polarization curves of all the specimens exhibit the similar shape. The polarization curves can be analyzed through cathodic Tafel extrapolation [28]. The electrochemical parameters, including corrosion potential (E_{corr}), corrosion current density (I_{corr}), and cathodic Tafel slope (b_c), are fitted and listed in Table 1. Figure 3 shows the distribution of corrosion current densities of array electrodes at the elbow. According to Figure 3, the corrosion current density increases from the outer wall to the inner wall of the elbow, and the maximum corrosion current density appears at the innermost side of the elbow. At the outermost side of the elbow, the corrosion current density decreases at first and then increases along the direction of fluid flow. However, the corrosion current density gradually increases from electrode 8 to electrode 11 and from electrode 12 to electrode 14 along the direction of fluid flow, which indicates the corrosion rate gradually increases from electrode 8 to electrode 11 and from electrode 12 to electrode 14 along the direction of fluid flow, respectively.

Table 1: The fitting parameters of polarization curves of array electrodes under fluid flow condition with the flow velocity of 5.31 m/s

E (electrode)	E_{corr} (mV(SCE))	B_c (mV/decade)	Corrosion current density (A/cm²)
1	−1524	−255.7	9.02×10^{-5}
2	−1562	−245.7	8.66×10^{-5}
3	−1551	−259.2	9.17×10^{-5}
4	−1555	−260.8	9.17×10^{-5}
5	−1572	−253.9	9.38×10^{-5}
6	−1607	−258.7	9.94×10^{-5}
7	−1603	−274.4	1.22×10^{-4}
8	−1531	−277.4	9.63×10^{-5}
9	−1521	−281.2	9.78×10^{-5}
10	−1566	−273.6	9.89×10^{-5}
11	−1564	−273.3	9.99×10^{-5}
12	−1543	−265.9	1.00×10^{-4}
13	−1542	−265.9	1.01×10^{-4}
14	−1556	−259.8	1.05×10^{-4}
15	−1537	−278.0	1.15×10^{-4}

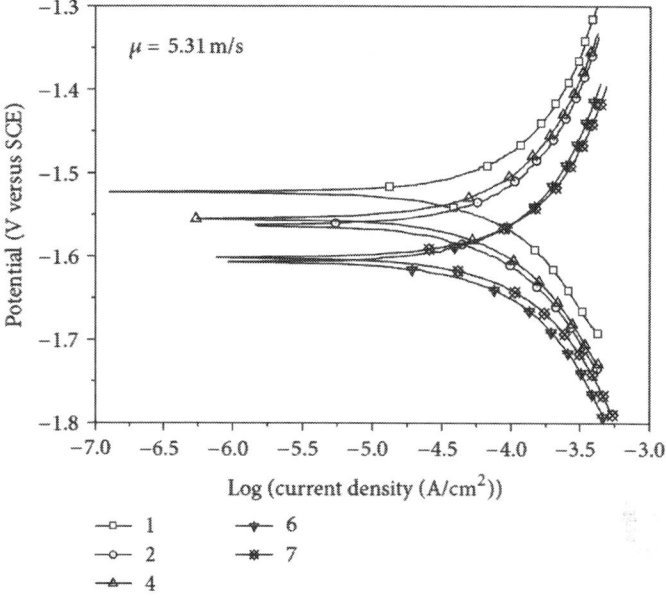

(a)

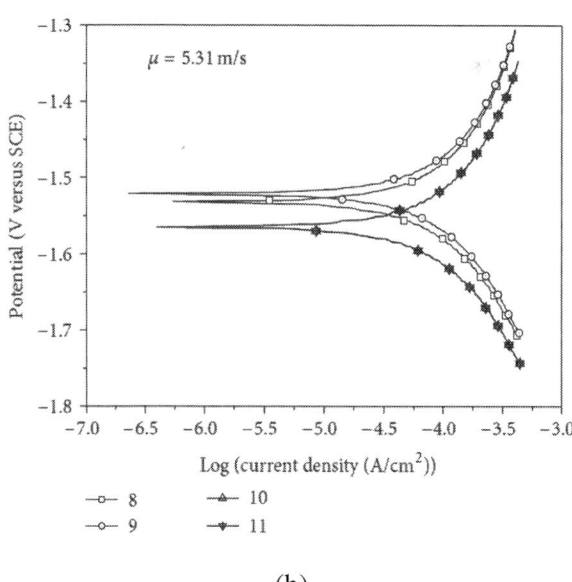

(b)

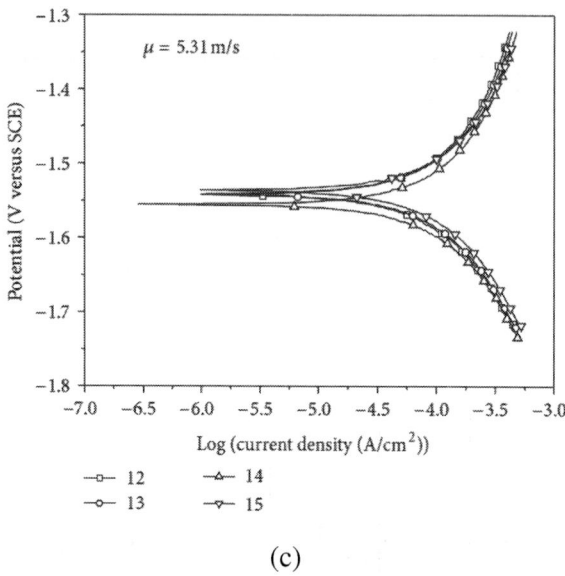

(c)

Figure 2: Potentiodynamic polarization curves of array electrodes under fluid flow condition with the flow velocity of 5.31 m/s.

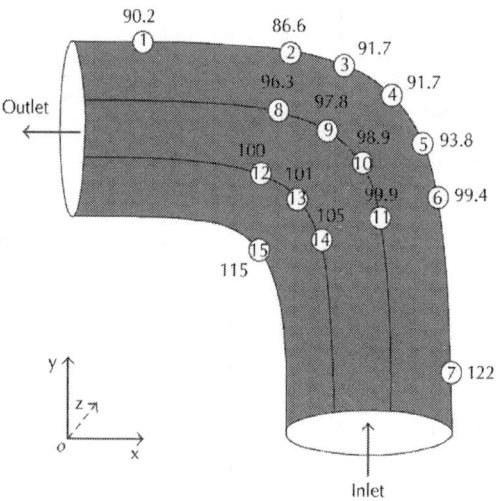

Figure 3: Distribution of corrosion current densities of array electrodes at test section under fluid flow condition with the flow velocity of 5.31 m/s (unit: $\mu A/cm^2$).

CFD Simulation

Figure 4 shows the three-dimensional distributions of fluid flow velocity and shear stress at the elbow with a flow velocity of 5.31 m/s. Three-dimensional distributions of fluid flow velocity and shear stress along the elbow are symmetrical with respect to the central plane of pipe. From the distribution of fluid flow velocity at the elbow (Figure 4(a)), flow velocity decreases from the innermost side to the outermost side, with the highest flow velocity about 6.88 m/s and the lowest flow velocity about 3.26 m/s. At the outermost side of the elbow, flow velocity decreases at first and then increases along the direction of fluid flow. In the straight section before the outlet of the elbow, flow velocity increases from the innermost side to the outermost side, with the highest flow velocity about 5.79 m/s and the lowest flow velocity about 3.26 m/s. From the distribution of shear stress at the elbow (Figure 4(b)), the shear stress also decreases from the innermost wall to the outermost wall, with the largest shear stress about 80.7 Pa and the smallest shear stress about 34.0 Pa. At the outermost side of the elbow, the shear stress decreases at first and then increases along the direction of fluid flow. In the straight section before the inlet of the elbow, the distribution of shear stress is uniform about 55.2 Pa. Based on above analysis, a conclusion could be drawn that the higher flow velocity corresponds to the larger shear stress and the lower flow velocity corresponds to the smaller shear stress at the elbow.

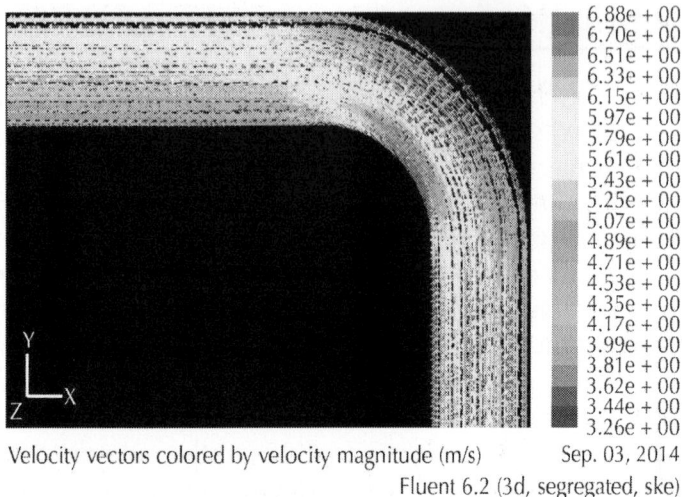

Figure 4: The three-dimensional distributions of fluid flow velocity and shear stress at elbow with the flow velocity of 5.31 m/s: (a) the distribution of fluid flow velocity and (b) the distribution of shear stress.

Surface Morphology Analysis after FAC Test

Figure 5 shows the surface morphologies of the electrode after corrosion under static state condition for 4 h. It is obviously seen from Figure 5 that there is a compact corrosion product film on the electrode surface. Figure 6 shows the surface morphologies of representative AZ91D magnesium alloy array electrodes after FAC test for 4 h. It is seen from Figure 6 that there are obvious cutting tracks on the surface of these electrodes. Furthermore, the scale of the corrosion product films becomes more and more greater from the inside to the outside of the elbow. This can be explained that the removal of corrosion products on the electrode surface is easy with the high flow velocity and the large shear stress, but it is difficult for corrosion products to remove with the low flow velocity and the small shear stress.

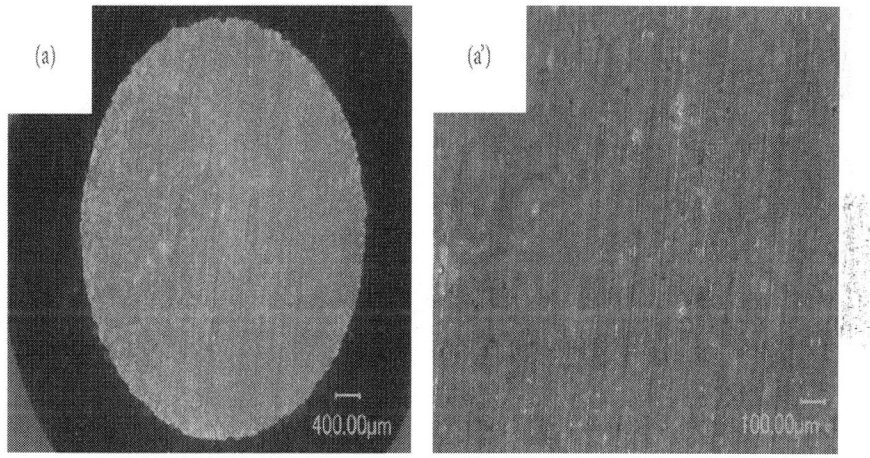

Figure 5: The surface morphologies of AZ91D after static state corrosion test for 4 h.

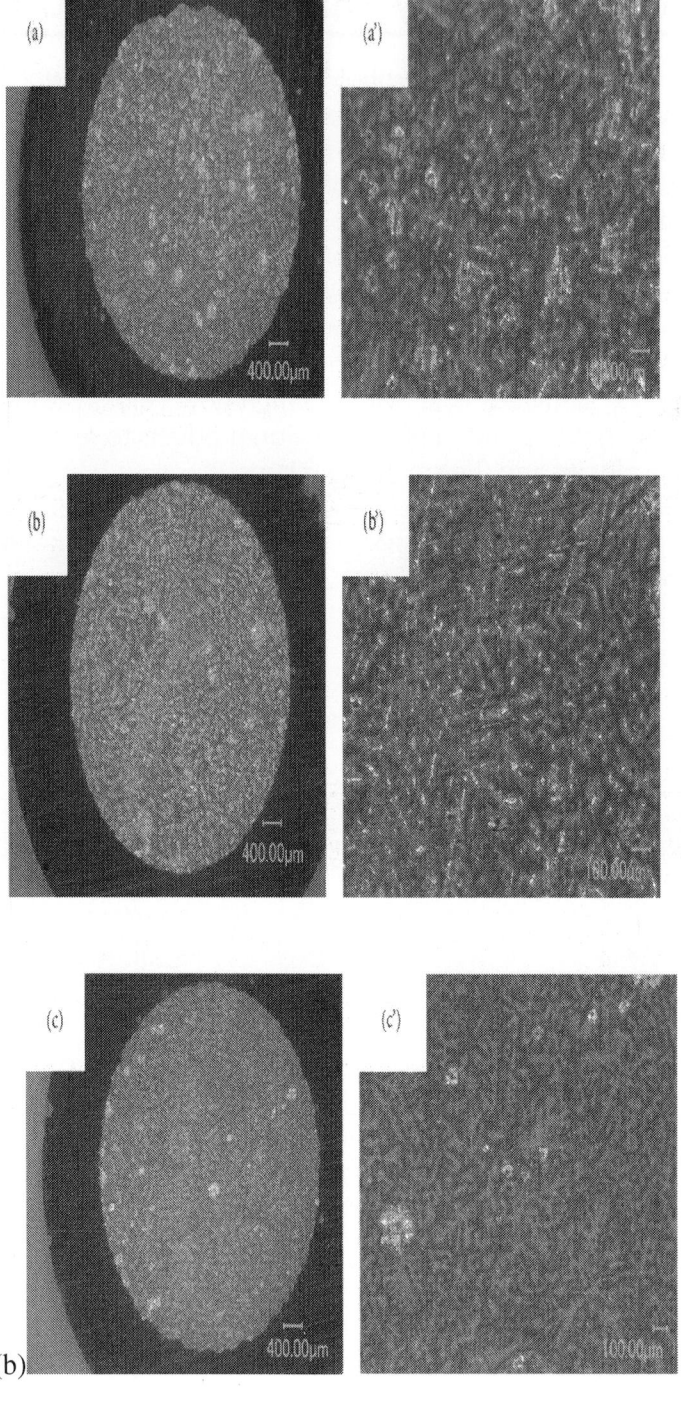

(b)

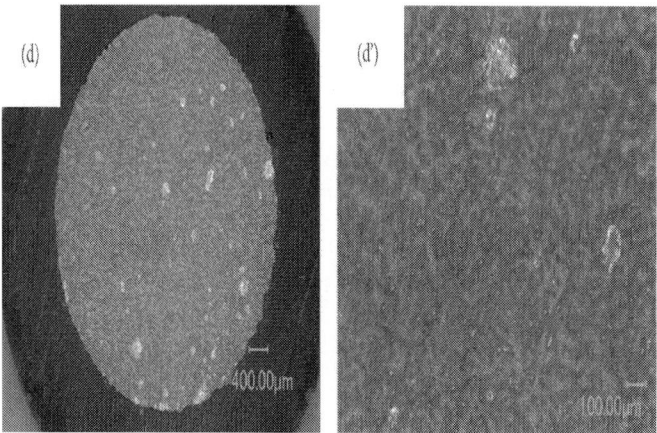

Figure 6: The surface morphologies of representative electrodes after FAC test for 4 h: (a) electrode 4, (b) electrode 10, (c) electrode 13, and (d) electrode 15.

DISCUSSION

The corrosion of magnesium alloys in chloride-containing solution should include anodic dissolution of magnesium and cathodic hydrogen evolution [12], which can always be expressed as the following reaction:

$$Mg \longrightarrow Mg^+ + e^- \quad \text{or} \quad Mg^+ \longrightarrow Mg^{2+} + e^-$$

$$\text{(1)}$$

$$2H_2O + 2e^- \longrightarrow 2OH^- + H_2 \quad \text{or} \quad 2H^+ + 2e^- \longrightarrow H_2$$

$$\text{(2)}$$

Although the corrosion mechanism of magnesium alloys may involve many complicated intermediate steps [5], the overall corrosion reaction can always be expressed as the following reaction:

$$Mg + 2H_2O \longrightarrow Mg(OH)_2 + H_2$$

$$\text{(3)}$$

Therefore, a corrosion product film consisting of $Mg(OH)_2$ can be formed on magnesium alloy surface.

Potentiodynamic polarization measurement demonstrates that

the electrochemical activity and corrosion behaviour of AZ91D magnesium alloy array electrodes are different at different locations of the elbow in circulating ASTM D1384-87 solution, as indicated in Figure 2. According to Figure 2, AZ91D magnesium alloy is in an active dissolution state in the flowing solution, which can be attributed that fluid flow enhances the convection and diffusion of reactive species, accelerating the corrosion electrochemical reactions. Simultaneously, the severity of the erosive/abrasive attack to AZ91D magnesium alloy electrodes increases with increasing fluid flow velocity and shear stress, resulting in the more active electrode state and thus increasing corrosion rate of AZ91D magnesium alloy.

In this work, the CFD simulation indicates that there are quite different fluid flow velocities and shear stress at different locations along the elbow due to the geometrical change. The maximum corrosion rate appears at the innermost wall of the elbow with the maximum flow velocity and shear stress. The corrosion rate decreases from the innermost wall to the outermost wall of the elbow, with the higher corrosion rate corresponding to the higher flow velocity and the larger shear stress at the elbow. In the outermost wall of the elbow, the corrosion rate decreases at first and then increases along the direction of fluid flow, with the same variation for fluid flow velocity and shear stress. By comparing the CFD simulation results to the experimental results, it is demonstrated that fluid hydrodynamics plays a significant role in the FAC of AZ91D magnesium alloy. This can be explained by the fact that the velocity gradient near the wall and shear stress on the electrode surface are high under the high flow velocity condition. Then, the mass transfer rate is higher at the inner wall than that at the outer wall which is caused by the hydrodynamics change due to the geometrical configuration and orientation at the elbow [13]. Therefore, the comparison of electrochemical measurement with CFD simulation shows that a highly active AZ91D magnesium alloy electrode with a high dissolution rate is associated with a high flow velocity and a large shear stress at the elbow.

Additionally, the surface morphologies show that the scale of the corrosion product films at the inner wall is smaller than that at the outer wall at the elbow. At the inner wall, the removal of corrosion products on the electrode surface is easy due to the high flow velocity and the large shear stress, accompanied with the increase of corrosion rate, while the corrosion products are relatively difficult to remove at the outer wall due to the low flow velocity and the small shear stress, resulting in a relatively low corrosion rate. Apparently, fluid hydrodynamics plays a significant role in the FAC of AZ91D magnesium alloy.

CONCLUSIONS

Based on array electrode technology, FAC of AZ91D magnesium alloy at an elbow was investigated in loop system by potentiodynamic polarization, computational fluid dynamics simulation, and surface analysis. It is demonstrated that fluid hydrodynamics plays a significant role in the FAC of AZ91D magnesium alloy, and the distribution of the corrosion rates is in good accordance with the distributions of fluid flow velocity and shear stress at the elbow. The corrosion rate increases from the outer wall to the inner wall of the elbow, with the higher corrosion rate corresponding to the higher flow velocity and the larger shear stress at the elbow. In the outermost wall of the elbow, the corrosion rate decreases at first and then increases along the direction of fluid flow, with the same variation for fluid flow velocity and shear stress. The maximum corrosion rate appears at the innermost wall of the elbow, the location with the maximum flow velocity and shear stress.

ACKNOWLEDGMENTS

The authors acknowledge the financial support of the National Natural Science Foundation of China (no. 51401151) and the Postdoctoral Science Foundation of China (no. 2012M511207). The authors also acknowledge the support of the Science Research Foundation of Wuhan Institute Technology (no. K201446).

REFERENCES

1.   G. Song, A. L. Bowles, and D. H. St John, "Corrosion resistance of aged die cast magnesium alloy AZ91D," Materials Science and Engineering A, vol. 366, no. 2, pp. 74–86, 2004.

2.   T. Zhang, X. Liu, Y. Shao, G. Meng, and F. Wang, "Electrochemical

noise analysis on the pit corrosion susceptibility of Mg–10Gd–2Y–0.5Zr, AZ91D alloy and pure magnesium using stochastic model," Corrosion Science, vol. 50, pp. 3500–3507, 2008.

3.   A. M. Fekry and M. Z. Fatayerji, "Electrochemical corrosion behavior of AZ91D alloy in ethylene glycol," Electrochimica Acta, vol. 54, no. 26, pp. 6522–6528, 2009.

4.   J. H. Nordlien, S. Ono, N. Masuko, and K. Nisancioglu, "Morphology and structure of oxide films formed on magnesium by exposure to air and water," Journal of the Electrochemical Society, vol. 142, no. 10, pp. 3320–3322, 1995.

5.   G. Song, A. Atrens, X. Wu, and B. Zhang, "Corrosion behaviour of AZ21, AZ501 and AZ91 in sodium chloride," Corrosion Science, vol. 40, no. 10, pp. 1769–1791, 1998.

6.   G. Baril, C. Blanc, and N. Pebere, "AC impedance spectroscopy in characterizing time-dependent corrosion of AZ91 and AM50 magnesium alloys characterization with respect to their microstructures," Journal of the Electrochemical Society, vol. 148, no. 12, pp. B489–B496, 2001.

7.   P. Schmutz, V. Guillaumin, R. S. Lillard, J. A. Lillard, and G. S. Frankel, "Influence of dichromate ions on corrosion processes on pure magnesium," Journal of the Electrochemical Society, vol. 150, no. 4, pp. B99–B110, 2003.

8.   N. LeBozec, M. Jönsson, and D. Thierry, "Atmospheric corrosion of magnesium alloys: influence of temperature, relative humidity, and chloride deposition," Corrosion, vol. 60, no. 4, pp. 356–361, 2004.

9.   J. Chen, J. Wang, E. Han, J. Dong, and W. Ke, "AC impedance spectroscopy study of the corrosion behavior of an AZ91 magnesium alloy in 0.1 M sodium sulfate solution," Electrochimica Acta, vol. 52, no. 9, pp. 3299–3309, 2007.

10. M. Ö. Öteyaka, E. Ghali, and R. Tremblay, "Corrosion behaviour of AZ and ZA magnesium alloys in alkaline chloride media," International Journal of Corrosion, vol. 2012, Article ID 452631, 10 pages, 2012. ·

11. S. A. Salman, R. Ichino, and M. Okido, "A comparative electrochemical study of AZ31 and AZ91 magnesium alloy," International Journal of Corrosion, vol. 2010, Article ID 412129,

7 pages, 2010.

12. G. L. Song and M. Liu, "The effect of surface pretreatment on the corrosion performance of Electroless E-coating coated AZ31," Corrosion Science, vol. 62, pp. 61–72, 2012.

13. G. A. Zhang, L. Zeng, H. L. Huang, and X. P. Guo, "A study of flow accelerated corrosion at elbow of carbon steel pipeline by array electrode and computational fluid dynamics simulation," Corrosion Science, vol. 77, pp. 334–341, 2013. ·

14. W. S. Miller, L. Zhuang, J. Bottema et al., "Recent development in aluminium alloys for the automotive industry," Materials Science and Engineering A, vol. 280, no. 1, pp. 37–49, 2000.

15. T. J. Harvey, J. A. Wharton, and R. J. K. Wood, "Development of synergy model for erosion-corrosion of carbon steel in a slurry pot," Tribology—Materials, Surfaces and Interfaces, vol. 1, no. 1, pp. 33–47, 2007.

16. S. S. Rajahram, T. J. Harvey, and R. J. K. Wood, "Erosion-corrosion resistance of engineering materials in various test conditions," Wear, vol. 267, no. 1–4, pp. 244–254, 2009.

17. S. S. Rajahramn, T. J. Harvey, and R. J. K. Wood, "Electrochemical investigation of erosion-corrosion using a slurry pot erosion tester," Tribology International, vol. 44, no. 3, pp. 232–240, 2011.

18. D. López, N. Alonso Falleiros, and A. Paulo Tschiptschin, "Effect of nitrogen on the corrosionerosion synergism in an austenitic stainless steel," Tribology International, vol. 44, no. 5, pp. 610–616, 2011.

19. R. C. Barik, J. A. Wharton, R. J. K. Wood, and K. R. Stokes, "Electro-mechanical interactions during erosion–corrosion," Wear, vol. 267, no. 11, pp. 1900–1908, 2009. ·

20. M. M. Stack and N. Pungwiwat, "Particulate erosion-corrosion of Al in aqueous conditions: some perspectives on pH effects on the erosion-corrosion map," Tribology International, vol. 35, no. 10, pp. 651–660, 2002.

21. G. T. Burstein and K. Sasaki, "Effect of impact angle on the slurry erosion-corrosion of 304L stainless steel," Wear, vol. 240, no. 1-2, pp. 80–94, 2000.

22. A. Neville and C. Wang, "Erosion-corrosion mitigation by

corrosion inhibitors—an assessment of mechanisms," Wear, vol. 267, no. 1–4, pp. 195–203, 2009.

23.   M. M. Stack and G. H. Abdulrahman, "Mapping erosion-corrosion of carbon steel in oil exploration conditions: some new approaches to characterizing mechanisms and synergies," Tribology International, vol. 43, no. 7, pp. 1268–1277, 2010.

24.   R. C. Barik, J. A. Wharton, R. J. K. Wood, K. S. Tan, and K. R. Stokes, "Erosion and erosion-corrosion performance of cast and thermally sprayed nickel-aluminium bronze," Wear, vol. 259, no. 1–6, pp. 230–242, 2005.

25.   R. Malka, S. Nešić, and D. A. Gulino, "Erosion-corrosion and synergistic effects in disturbed liquid-particle flow," Wear, vol. 262, no. 7-8, pp. 791–799, 2007.

26.   R. O. Rihan and S. Nesic, "Erosion–corrosion of mild steel in hot caustic. Part I: NaOH solution," Corrosion Science, vol. 48, no. 9, pp. 2633–2659, 2006. ·

27.   M. El-Gammal, H. Mazhar, J. S. Cotton, C. Shefski, J. Pietralik, and C. Y. Ching, "The hydrodynamic effects of single-phase flow on flow accelerated corrosion in a 90-degree elbow," Nuclear Engineering and Design, vol. 240, no. 6, pp. 1589–1598, 2010.

28.   J. Hu, D. Huang, G. Song, and X. Guo, "The synergistic inhibition effect of organic silicate and inorganic Zn salt on corrosion of Mg-10Gd-3Y magnesium alloy," Corrosion Science, vol. 53, no. 12, pp. 4093–4101, 2011.

Chapter 4

Effect of Corrosion Inhibitors in Limestone Cement

Evgenia Zacharopoulou, Aggeliki Zacharopoulou,
Atteyeh Sayedalhosseini, and George Batis,
Sotirios Tsivilis

Department of Materials Science and Engineering, School of Chemical
Engineering, National Technical University of Athens, Athens, Greece

ABSTRACT

In this paper examines the improving durability of different limestone
cement and effects of the use of corrosion inhibitor. The target is to
experimentally investigate the effect of different types of cement in
corrosion of reinforcement in presents of corrosion inhibitors and
without it. Three types of cement have been used: CEM II, LC1 and
LC2. For this purpose constructed mortar specimens, containing 4
reinforcements, with or without corrosion inhibitors for each group,
these exhibited to partial immersion in sodium chloride in 3.5% w.t
NaCl solution. The methods, with which the corrosion of reinforcement
in concrete was tested, were measurements of corrosion potential,
corrosion current and mass loss of reinforcement. The mortars with

CEM II cement have better durability than that with limestone cement. The use of VpCI, Cyclohexylammonium benzoate, improves the corrosion protection of mortars with CEM II cement upper 50%. On the other hand, the addition of VpCI, Cyclohexylammonium benzoate, improves the corrosion protection of mortars with limestone cement 30% or lower.

INTRODUCTION

The cement industry continues to introduce more sustainable practices and products for constructing and maintaining our concrete infrastructure and buildings. That sustainable development focus, proposed implementation of more restrictive environmental regulations on cement manufacturing, and a legislation of potential global climate change has prompted the US cement industry to propose provisions for Portland-limestone cements within specifications ASTM C595 [1] and AASHTO M240 [2]. Portland-limestone cements are in common use around the world.

Limestone that is provided from the European standard EN 197-1 (CEN 2000) allows cements to contain limestone in three different dosage levels. CEM I, "Portland cement", may contain up to 5% minor additional constituents, of which limestone is one possible material. CEM II/A-L and CEM II/B-L, both called "Portland limestone cement", contain 6% to 20% and 21% to 35% ground limestone, respectively. Roughly 19% of all cement sold in Europe contains between 6% and 35% limestone [3]. The requirements were specified for limestone use pertaining to effects on performance only.

In most early research it was believed that limestone acted as inert filler; however more recent researches have shown that limestone participates to some extent in hydration reactions. In addition, fine limestone particles may promote silicate hydration by providing nucleation sites for C-S-H precipitation.

Calcium carbonate has been reported to react with the tricalcium aluminate to form high and low forms of carboaluminates [4].

Tsivilis [5] found that the addition of limestone as an intergrinding material increased the reactivity of the clinker. Campiteli and Florindo [6] found that the addition of limestone decreased the optimum SO_3

content. Production of CH appears to increase at early ages, which was attributed in part to the dissolution of limestone and in part to the role of the limestone in acting as a nucleation site [7].

Permeability is the key to the durability of a porous material in all but the most protected environments. With the exceptions of abrasion and erosion, deterioration mechanisms involve the ingress of water and/or other harmful species (oxygen, carbon dioxide, chlorine ions, sulfate ions, acids, etc.). Corrosion requires water and oxygen, and is catalyzed by chlorine ions.

In a related study, Tsivilis [8] produced concretes with five cements with limestone contents ranging from 0% to 35%, and conducted the "Rapid Chloride Permeability Test" (RCPT) (ASTM C1202) after 28 days of moist curing. Table 1 shows details of the cements and concrete together with the results of the RCPT. Addition of lime over 10% increases the value of RCPT. The results show little impact due to the increase of limestone content up to 15 % to 20%. The mix with 35% limestone had a higher RCPT value despite being cast with a lower w/cm, indicating that permeability increased at this level of limestone [5].

This paper examines the improving durability of two different limestone cements compared with CEM II cement and the protective effects of the use of corrosion inhibitor. The methods, with which was tested the corrosion of reinforcement in concrete, were: measurements of corrosion potential, corrosion current and electrochemical calculated mass loss of reinforcement.

ANALYTICAL INVESTIGATION

Materials

Cements

In the present experiment 3 type of cement were used CEMII, LC1 and LC2. Portland limestone cements, contain 15%, 35% w/w limestone, were produced by intergrinding of clinker, limestone and gypsum in a pilot plant ball mill of 5 kg capacity. Preliminary tests with varying grinding time have been done in order to produce cements of appropriate compressive strength. The cements, LC1 and LC2, contain 15%, 35% limestone, respectively. The composition of the cement, as well as their 28 days compressive strength and specific surface are shown in Table 1 [9].

The chemical and mineralogical composition of clinker is shown in Table 2.

The chemical composition of the CEMII cement is shown in Table 3 [10].

Portland cement clinker of industrial origin and limestone of high calcite content ($CaCO_3$: 97.5%) were used. The chemical composition of the above materials is present in Tables 3 and 4 respectively. The used clinker has a moderate C_3A.

Table 1: Characteristic of the LC1 and LC2 cements

Code	Cement composition	28-days compressive strength (Mpa)	Specific surface (cm² /g)
LC1	Clinker: 85% w/w, limestone: 15% w/w	41.3	3980
LC2	Clinker: 65% w/w, limestone: 35% w/w	32.4	5040

Table 2: Chemical and mineralogical composition of clinker

Chemical composition (%)		Mineralogical composition (%)	
SiO_2	21.92	C_3S	48.4
Al_2O_3	5.68	C_2S	26.3
Fe_2O_3	3.29	C_3A	9.5

CaO	63.35	C_4AF	10.00
MgO	1.44		
K_2O	1.32	Moduli	
Na_2O	0.84	Lime Saturation Factor (LSF)	95.7
SO_3	1.25	Silica Ratio (SR)	2.5
LO1	0.91	Alumina Ratio (AR)	1.43
fCa0	1.15	Hydraulic Modulus (HM)	2.18

Aggregates

The use of aggregates should be according to EN12620 for normal and heavyweight aggregates and according to EN13055-1 for lightweight aggregates. In this sand have been used aggregate, which confirm all required terms and conditions.

Reinforcement

Concrete reinforcement steel should be protected against corrosion, both before it is incorporated into concrete, and after. During its placement into the final position, steel should be relieved of all visible scaling alterations, or unwanted deformations and damage, which, decides other things, speed up the effects of corrosion.

In this experiment we use steels type of B500C ELOT. The marking for identifying the quality by a grad of concrete reinforcement steel is done with a different configuration of the transverse ribs on the surface of the bar.

Steels with ribs are characterized by their surface geometry, which dictates their adhesion to the concrete. Concrete reinforcement steels with ribs have at least two rows of parallel transverse ribs uniformly distributed on each side of the steels surface and at equal distance throughout both rows. Longitudinal ribs may be added, but are not mandatory.

The tensile strength limits that the mechanical properties of concrete reinforcement steels must meet are given in Table 5.

The values of yield strength f_y and f_i are calculated according to the nominal cross-section.

The chemical composition and production methods are display in Table 6.

Table 3: Chemical composition of CEMII cement

SiO_2	AL_2O_3	Fe_2O_3	Cao	Mgo	K_2O	Na_2O	SO_3	CaO_f	LOI	Specific surface (cm^2/g)
27.38	9.1	5.65	45.39	2.73	0.94	0.56	2.71	2.67	5.04	3900

Table 4: Chemical analysis of limestone (% w/w)

SiO_2	AL_2O_3	Fe_2O_3	Cao	Mgo	K_2O	Na_2O	SO_3	LOI	Total
0.57	0.33	0.19	54.60	1.64	0.04	-	0.02	42.76	100.15

Table 5: The tensile strength limits of the mechanical properties of steel according to ELOT 1421-2 and ELOT 1421-3 (typical values)

Property	Technical Quality Grade
	500C
Yield strength, f_y (MPa)	≥ 500
Ratio of the actual to the nominal value of yield strength $f_{y,act}/f_y$ nom	≤ 1.25
Ratio of tensile to yield strength f_t/f_y	$\geq 1.15 \leq 1.35$
Total Strain (elongation) under maximum load u (%)	≥ 7.5

Table 6: Chemical composition, production methods

Steel Grade	Common chemical composition (% of weight)				Production Method
	C	Mn	Si	S	
B500C	0.20 - 0.22	0.90 - 1.20	0.15 - 0.30	0.03 - 0.05	HRT (Hot Rolled Tempered steel)

Corrosion Inhibitor

The corrosion inhibitor used is Cyclohexylammonium benzoate, as admixtures. Cyclohexylammonium benzoate is a VpCI (Vapor Corrosion inhibitor) volatile low pressure corrosion inhibitor and multifunctional corrosion inhibitor. Protects rebars electrochemically through anodic or cathodic reaction as contact inhibition and diffusion of molecules (potential migration molecules). The nature of the product is to exploit whatever substrate absorption, creating diffusion through the porous concrete, approaching the reinforcing steel (structural steel elements) in order to protect them from further corrosion, extending both the length and the operating limits of construction. The addition of VpCI is 0.4/100g cement. The structural formula Cyclohexylammonium benzoate is the following [11]:

$$CH_2 - CH_2 - CH - NH_3^+$$
$$\mid \qquad\qquad \mid$$
$$CH_2 - CH_2 - CH_2 - O_2CC_6H_5$$

Further technical characteristics and physical properties are presented in Table 7.

Table 7: Technical characteristics and physical properties of the corrosion inhibitor

Display-Painting	White milky liquid
Waiting time between coats 8 - 24 hours	8 - 24 hours
Specific gravity	1.03 ± 0.01 kg/lt
Application Temperature	Minimum +1°C (ambient and substrate)
Consumption	1 2 coats x ~ 0.0150 kg/m²
Grade pH	8.9 - 9.4

Methods

As part of the experimental work, electrochemical studies as monitoring the corrosion potentials, linear polarization and mass loss of rebars were carried out as described below.

Monitoring the Corrosion Potentials

The corrosion trend of the samples was estimated by monitoring the corrosion potential versus exposure time. The corrosion potential of steels was measured according to ASTM C876-87, using a Ag/AgCl/KCl (3M) electrode as a reference electrode which was placed in 3.5% NaCl solution. A voltmeter is also needed.

As it's referred in ASTM C 876-87:

- Potentials less negative than −0.2 volts generally indicate 90% higher probability of no corrosion taking place at the time of measurement.
- Potentials in the middle of −0.2 to 0.35 volts are inconclusive.
- Potentials greater than −0.35 volts generally indicated, 90% or higher probability of active corrosion in the area at the time of testing.
- Positive potentials, if obtained, generally indicate insufficient moisture in the concrete and should not be considered valid. However, stray DC currents may also cause potential measurements and therefore careful review analysis of the obtained data is required.

Linear Polarization

The linear polarization technique requires us to polarize the steel with an electric current and monitor its effect on the half cell potential. It is carried out with a sophisticated development of the half cell incorporating an auxiliary electrode and a variable low voltage DC power supply. The half cell potential is measured and then a small current is passed from the auxiliary electrode to the reinforcement. The change in the half cell potential is simply related to the corrosion current by the equation:

$$I_{corr} = B/R_p$$

where B is a constant (in concrete 26 to 52 mV) and it depend on type of cement, water cement ratio and the passivity or active condition of the steel. For steel in passivity condition B = 52 mV and for steel in active condition B = 26 mV. In this present thesis, assuming that steel was in active condition, B = 26 mV. R_p is the polarization resistance (in ohms):

$$R_p = (\text{change in potential})/(\text{applied current})$$

R_p gives the technique its alternative name of polarization resistance. The change in potential must be kept to less than 20 mV or so for the equation to be valid and remain linear.

The change in potential must be kept to less than 20 mV or so for the equation to be valid and remain linear. The "iR drop" must also be removed. This is the voltage that exists because a current is flowing through concrete that has an electrical resistance. This is also referred to as the solution resistance. This means that the current is usually switched off during the measurement process so that the potential without the iR drop is measured [12]. Corrosion is an electrochemical process whereby the amount of corrosion is related to the electrical energy consumed, which is a function of voltage, amperage, and time interval. The amount of corrosion can be estimated using an equation based on Faraday's low:

$$\beta = t \cdot M \cdot I_{corr}/z \cdot F$$

where: t is time (sec), I_{corr} is corrosion current (Amperes), M is atomic weight of iron (55.847 g/mol), z is in charge (assumed 2 for $Fe \rightarrow Fe^{2+}+2e^-$) and F is Faraday's constant (96.487 Amp.sec).

EXPERIMENTAL PROCEDURE

During this project all Reinforcements were cleaned by immersing in acetone in order to removes grease and oils, immersing in solution HCl and corrosion inhibitors to remove rust from steel. Steel bars were sunk in deionized water, in Acetone and alcohol. Few seconds later they were weighted.

The specimens were prepared using three different cement types CEMII, LC1, LC2. In all specimens the aggregate used was the same. Reinforcing steel bars of steel type B500C and tap water were used. Mix proportions aggregate/water/cement was kept constant and equal to 3/1/0.5. Each specimen was cast into a prismatic mould (80 × 80 × 100 mm), where four identical steel bars (100 × 12 mm) was embedded in position shown in Figure 1.

Specimens were stored at ambient condition for 48 h, then cure in tap water for 7 days and then the part shown in Figure 1 was insulated with epoxy glue. Finally, all specimens were partially immersed up to 2/3 of their height in 3.5% NaCl.

MEASUREMENTS

Measurement of Corrosion Potential

Corrosion potentials were measured for a period of 305 days for specimens. Measurement of corrosion potentials for specimens started after passing 12 days from immersing in 3.5% NaCl. The measurements of corrosion potential for CEM II, LC1 and LC2 without corrosion inhibitor are shown in Figure 2.

Figure 2 shows the comparison of corrosion potential for CEM II, LC1 and LC2 without corrosion inhibitor. It is noticed that CEM II without corrosion inhibitor has better behavior than LC2 without corrosion inhibitor and LC2 without corrosion inhibitor has better behavior than LC1 without corrosion inhibitor.

The measurements of corrosion potential for CEM II, LC1 and LC2 with corrosion inhibitor are shown in Figure 3.

Figure 3 shows the comparison of corrosion potential for CEM II, LC 1 and LC 2 with corrosion inhibitor. It is noticed that CEM II with corrosion inhibitor has better behavior than LC2 with corrosion inhibitor and LC2 with corrosion inhibitor has better behavior than LC1 with corrosion inhibitor.

Electrochemical Calculated Mass Loss

The corrosion rate of the bars was determined by measuring their mass losses at predetermined exposure time intervals in the corrosive environments. In order to find the mass loss the polarization resistance (R_p) was measured and corrosion current (I_{corr}) and mass loss were calculated.

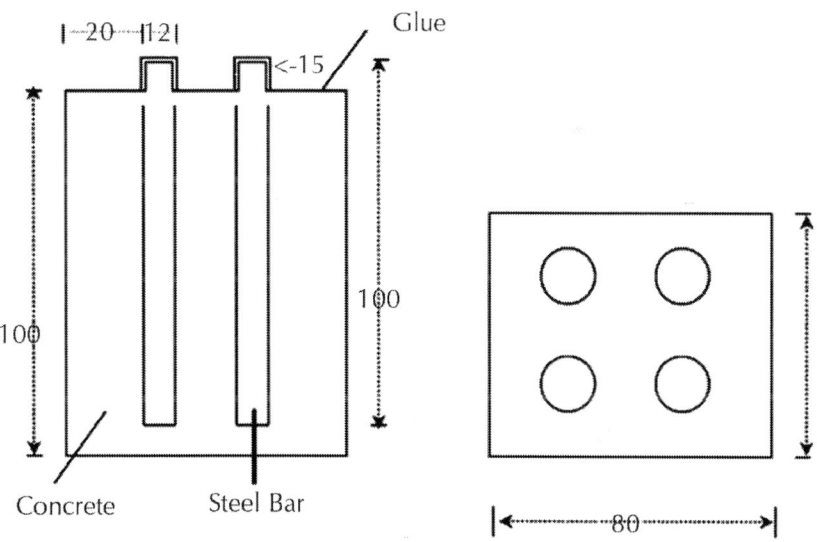

Figure 1: Schematic representation and dimensions (mm) of specimens.

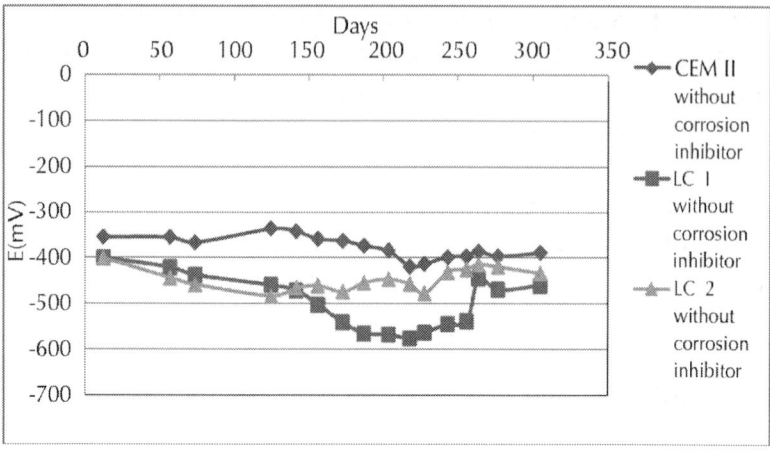

Figure 2: Figure of corrosion potentials for CEM II, LC1 and LC2 without corrosion inhibitor.

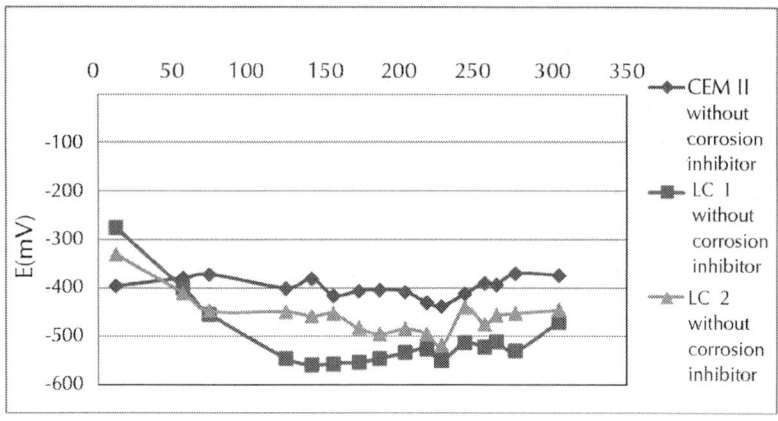

Figure 3: Figure of corrosion potentials for CEM II, LC1 and LC2 with corrosion inhibitor.

The measurements of mass loss for CEM II, LC1, LC2 without corrosion inhibitor are shown inFigure 4.

From Figure 4 is obvious that the durability of mortars with CEM II cement is better than LC1 and LC2.

The measurements of mass loss for CEM II with and without corrosion inhibitor are shown in Figure 5.

Figure 5 shows the comparison of mass loss of CEM II with and without corrosion inhibitor. It is noticed that the CEM II without corrosion inhibitor appears greater mass loss than the CEM II with corrosion inhibitor. It is obvious that the corrosion inhibitor protects the rebars from the corrosion.

The measurements of mass loss for LC1 with and without corrosion inhibitor are shown in Figure 6. It is known that the concrete or mortars with limestone cement have lower durability as concrete or mortars with CEM I and CEM II cement [13].

Figure 6 shows the comparison of mass loss of LC1 with and without corrosion inhibitor. It is important to mention that the mass loss of LC1 with and without corrosion inhibitor is approximately the same. From the measurements of mass loss of LC1 with and without corrosion inhibitor, there is no significant difference.

The measurements of mass loss for LC2 with and without corrosion inhibitor are shown in Figure 7.

Figure 7 shows the comparison of mass loss of LC2 with and without corrosion inhibitor. It is noticed that the LC2 without corrosion inhibitor appears greater mass loss than the CEM II with corrosion inhibitor. It is obvious that the corrosion inhibitor protects the rebars from the corrosion.

The measurements of mass loss for CEM II, LC1, LC2 with corrosion inhibitor are shown in Figure 8.

Figure 8 show that the protective action of corrosion inhibitor is greater in mortars with CEM II cement.

DISCUSSION

It is known that the concrete or mortars with limestone cement have lower durability as concrete or mortars with CEM I and CEM II cement [13]. It is also known that the addition of corrosion inhibitors such as calcium nitrite, N-N'-dimethylaminoethanol increases the protection of rebars of mortars or concrete from corrosion in chloride environment [6,10]. In this study is made an effort to promote the protection of rebars, of mortars with lime- stone cement, with VpCI Cyclohexylammonium benzoate in chloride contaminated environment.

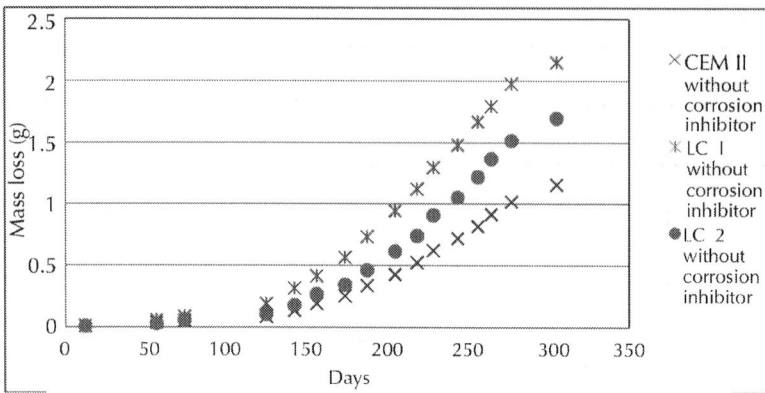

Figure 4: Figure of (electrochemical calculated) mass loss for CEM II, LC1, LC2 without corrosion inhibitor.

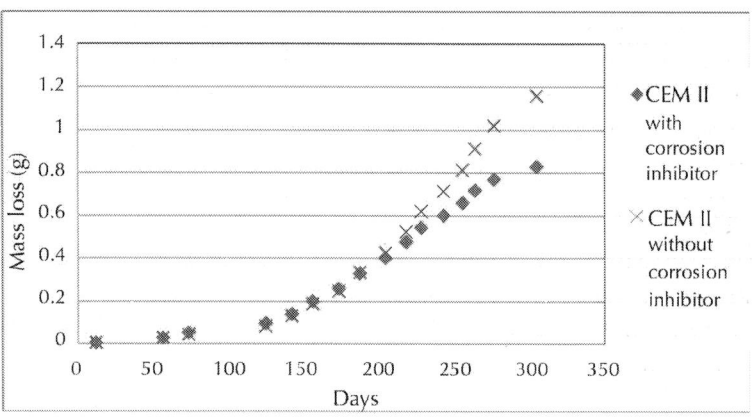

Figure 5: Figure of (electrochemical calculated) mass loss for CEM II with and without corrosion inhibitor.

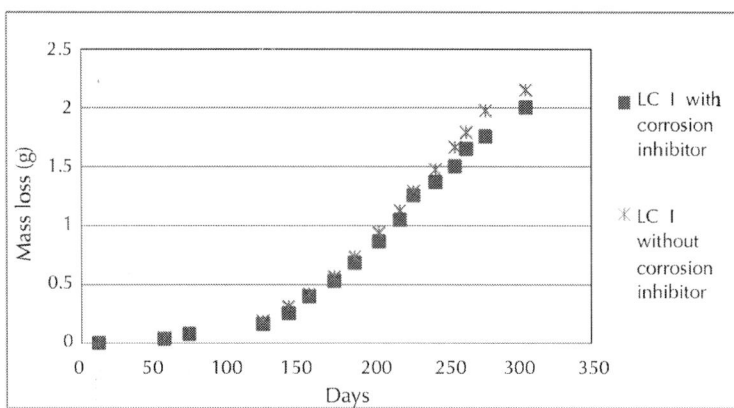

Figure 6: Diagram of (electrochemical calculated) mass loss for LC1 with and without corrosion inhibitor.

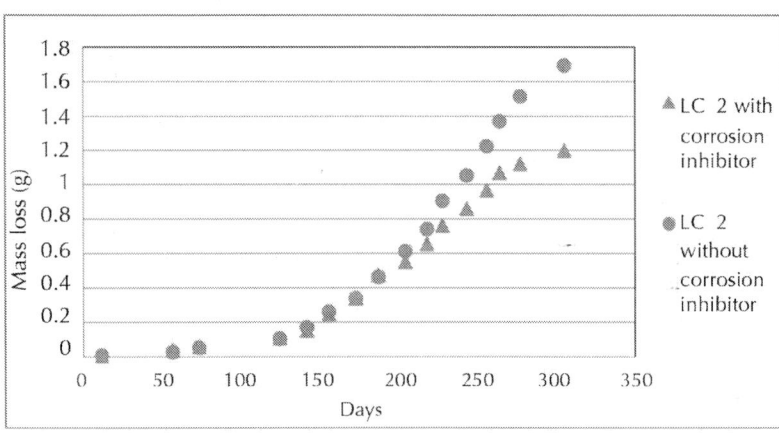

Figure 7: Diagram of (electrochemical calculated) mass loss for LC2 with and without corrosion inhibitor.

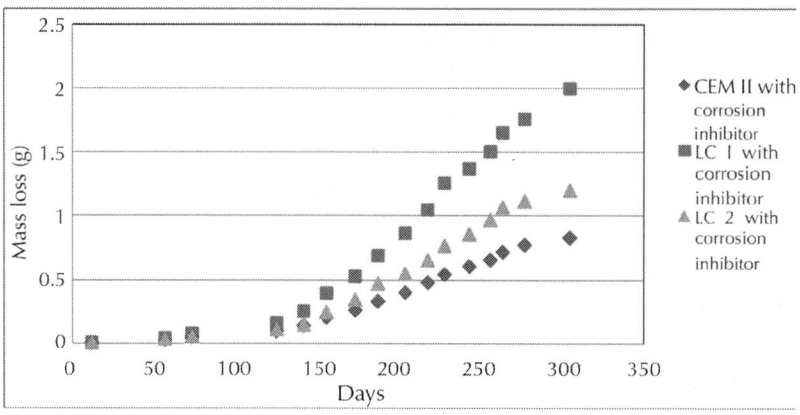

Figure 8: Figure of (electrochemical calculated) mass loss for CEM II, LC1, LC2 with corrosion inhibitor.

From the half-cell potential of rebars versus time (see Figures 2 and 3), it is no possible to establish a protective action of corrosion inhibitor. From the electrochemical calculated mass loss (see Figures 4-6) the VpCI Cyclohexylammonium benzoate has a protective action to corrosion of rebars after 200 days.

The addition of VpCI in mortar with CM II cement has corrosion protective effect about 52% (calculated for 300 days). On the same time, the LC2 mortar with limestone cement has corrosion protective effect of 30%. For the LC1 the protective effect is lower.

The corrosion protective effect of mortar with CEM II cement is compared with the general opinion that the corrosion inhibitors duplicate the technical service life of concrete [4,9,14]. On the other hand, the corrosion protective effect of mortars with limestone cement has lower protective effect. It is possible that the VpCI Cyclohexylammonium benzoate cannot be used in mortar and concrete with limestone cement.

The degreasing of corrosion protective effect, it is possible to explain through the observation, that the use of concrete with limestone cement is entrapped in large air voids that concrete with CEM I and II cement [12,15]. In this case the simultaneously use of VpCI and air-entraining can be improved the percentage of VpCI corrosion protection.

RESULTS

Based on the results of this experimental investigation under corrosive environment, the following conclusions are drawn:

- The mortars with CEM II cement have better durability than the limestone cements.
- The use of VpCI Cyclohexylammonium benzoate improves the corrosion protection of mortars with CEM II cement 52% (calculated for 300 days).
- The use of VpCI Cyclohexylammonium benzoate improves the corrosion protection of mortars with limestone cement LC2 (35% limestone) 30%.
- The use of VpCI Cyclohexylammonium benzoate improves the corrosion protection of mortars with limestone cement LC1 (15% limestone) only 10%.

REFERENCES

1. ASTM C595, "Standard Specification for Blended Cements."
2. AASHTO M240, "Standard Specification for Blended Cement."
3. V. C. Campiteli and M. C. Florindo, "The Influence of Limestone Additions on Optimum Sulfur Trioxide Content in Portland Cements," In: P. Klieger and R. D Hooton, Eds. Carbonate Additions to Cement, ASTM STP 1064, American Society for Testing and Materials, Philadelphia, 1990, pp. 30-40.
4. D. K. Yfantis, "Materials—Corrosion and Protection," Edition NTUA, Athens, 2000.
5. K. Sotiriadis, E. Nikolipoulou and S. Tsivilis, "Sulfate Resistance of Limestone Cement Concrete Exposed to Combined Chloride and Sulfate Environment at Low Temperature," Cement and Concrete Composites, Vol. 34, No. 8, 2012, pp. 903-910.
6. G. Batis, K. K. Sideris and P. Pantazopoulou, "Influence of Calcium Nitrite Inhibitor on the Durability of Mortars under Contaminated Chloride and Sulphate Environments," Anti-Corrosion Methods and Materials, Vol. 51 No. 2, 2004, pp. 112-120.http://dx.doi.org/10.1108/00035590410523201

7.   S. Tsivilis, E. Chaniotakis, G. Kakali and G. Batis, "An Analysis of the Properties of Portland Limestone Cements and Concrete," Cement and Concrete Composites, Vol. 24, No. 3, 2002, pp. 371-378.

8.   S. Tsivilis, G. Batis, E. Chaniotakis, G. Grigoriadis and D. Theodossis, "Properties and Behavior of Limestone Cement Concrete and Mortar," Cement and Concrete Research, Vol. 30, No. 10, 2000, pp. 1679-1683. http://dx.doi.org/10.1016/S0008-8846(00)00372-0

9.   M. C. Brown, "Assessment of Commercial Corrosion Inhibiting Admixtures for Reinforced Concrete," Virginia Polytechnic Institute, Blacksburg, 1999.

10. G. Batis, N. Kouloumbi and P. Pantazopoulou, "Protection of Reinforced Concrete by Coatings and Corrosion Inhibitors," Pigment and Resin Technology, Vol. 29, No. 3, 2000, pp. 159-163. http://dx.doi.org/10.1108/03699420010334312

11. "Carbonation of Concrete," Concrete Experts International, Denmark, 2002.

12. P. Turker and K. Erdo du, "Effects of Limestone Addition on Microstructure and Hydration of Cements," Proceedings of the Twenty Second International Conference on Cement Microscopy, Montreal, 29 April-4 May 2000.

13. R. D. Hooton, M. Nokken and M. D. A. T. Thomas, "Portland-Limestone Cement: State-of-the-Art Report and Gap Analysis for CSA A 3000," Report SN3053, Cement Association of Canada, Toronto, 2007, 60 p. http://www.cement.org/bookstore/results_quicksearch.asp?store=main&id=&cat2ID=3&searchterm=SN3053

14. B. P. John, "Corrosion of Steel in Concrete," E&FN Spon, London, 1997. http://dx.doi.org/10.4324/9780203414606

15. S. Tsivilis, E. Chaniotakis, G. Batis, C. Meletiou, V. Kasselouri, G. Kakali, A. Sakellariou, G. Pavlakis and C. Psimadas, "The Effect of Clinker and Lime Stone Quality on the Gas Permeability, Water Absorption and Pore Structure of Limestone Cement Concrete," Cement and Concrete Composites, Vol. 21, No. 2, 1999, pp. 139-146.

Chapter 5

Effect of Residual Stress on the Corrosion Behavior of Austenitic Stainless Steel

Osamu Takakuwa and Hitoshi Soyama

Department of Nanomechanics, Graduate School of Engineering, Tohoku University, Sendai, Japan

ABSTRACT

In this paper we demonstrate that the residual stress introduced by several different surface finishes affects the critical current density for passivation and the passive current density in the anodic polarization curve of austenitic stainless steel and that those critical current densities can be reduced by controlling the residual stress by applying a cavitating jet to the backs of specimens. The results show that the current density either increased or decreased depending on the surface finish, and that was decreased by introducing compressive residual stress for all surface finishes.

INTRODUCTION

The objective of this paper is to demonstrate the effect of residual stress on the corrosion behavior in austenitic stainless steel 316L. Residual stress is introduced into machine components and structures by surface finishing and/or heat treatments, and this affects the fatigue life and resistance to stress corrosion cracking [1] [2]. Several surface finishes such as a polisher and grinder are used for finishing the surface and removing the asperity generated by welding process at surface of components and structures. For improving the reliability of components and structures, mitigating the tensile residual stress is needed. The introduction of compressive residual stress in the surface layer can enhance the fatigue life and resistance to stress corrosion cracking, whereas tensile residual stress does the opposite [3] [4]. For this reason, peening techniques using the impact of shot [5], cavitation bubbles collapsing [6] or pulsed laser [7] have been developed in order to introduce compressive residual stress in practical applications. Since the residual stress at the surface is a significant factor, it is important to understand the effect of this on the corrosion behavior of materials, particularly for components and structures used in corrosive environments. Therefore, in this paper, we show the effect of the residual stress, not on stress corrosion cracking, but on the corrosion behavior. We do this by measuring the anodic polarization curve, and evaluating the critical current density for passivation and the passive current density while controlling the residual stress at the surface.

Improving the resistance of a material to corrosion is necessary to increase its life and reliability in a corrosive environment; consequently, several studies have been conducted on the corrosion behavior of metals. In particular, stainless steel has been developed for use in corrosive environments, and the corrosion resistance and electrochemical behavior have been widely studied [8] - [11]. The corrosion behavior can be evaluated from the polarization curve, which is obtained by measuring the current density as it varies with applied voltage in steel soaked in a corrosive solution. In this paper, the applied voltage was increased from the natural potential which is the equilibrium potential of the cathodic and anodic reactions. The current density required to generate a passivation film, i.e., the critical current density for passivation, and the current density required to maintain the passivation layer, i.e., the passive current density, were used as indices of corrosion behavior by reference to previous studies [10] [11] .

In the case of stainless steel, it is well-known that chromium reacts in preference to iron and that the progress of corrosion is suppressed by generating a passivation film. The corrosion behavior is affected by several factors such as the metallic structure, i.e., whether it is austenite, martensite or ferrite. The current density in austenite was increased by generating ferrite and carbide by a heating effect during the welding process [12]. In contrast, it was decreased by laser peening and shot peening with the introduction of compressive residual stress and nanocrystallization in 316L and 304 stainless steels [13] -[15] . It has been reported that the cold rolling ratio, which includes changes in the metallic structure and the residual stress, affects the corrosion behavior in 304 and 316L stainless steels [16] - [18]. Since there is a possibility that the residual stress in the surface layer affects the critical current density and the current density for passivation which are closely related to formation and retention of the passivation layer, it should be investigated. Several surface finishes are used at the end of process such as removing the roughness generated by welding process in the chemical and/or power plant. It affects the surface texture [19] and the sensitivity of stress corrosion cracking. A. Turnbull demonstrated that the surface machining and grinding affected the residual stress at surface layer, and consequently, it affected the stress corrosion cracking behavior [20].

In this paper, in order to demonstrate the effect of residual stress on the corrosion behavior in austenitic stain- less steel, specimens made of Japanese Industrial Standards JIS SUS316L were treated by several different surface finishes, and then subjected to a corrosion test using an electrochemical method. The residual stresses in the specimens were controlled by generating curvature in them by exposing the backs of them to a cavitating jet. The effect of residual stress on the current densities related to generating and maintaining a passivation film at the surface was investigated.

EXPERIMENTAL APPARATUS AND PROCEDURES

Figure 1 shows a schematic diagram of the apparatus used to measure the anodic polarization curve. This measurement was conducted based on the Japanese Industrial Standards JIS G 0579. The corrosive solution was a 5 percent by mass H_2SO_4 solution. The solutions were prepared by adding the requisite amount of ion-exchanged water, with the temperature maintained at 30 degrees Celsius in a temperature-controlled bath. The test electrode was the specimen. The counter electrode and reference electrode were a platinum electrode and a silver-silver chloride (Ag/AgCl) in saturated KCl electrode, respectively. Only an area of 24 mm diameter was exposed to the solution using the plastic holder shown in Figure 2. The standoff distance between the surface and the Luggin probe is 1 mm. Before conducting the corrosion test, the test solution was deaerated by flowing Argon gas through it at greater than 100 ml/min for more than half an hour, and the surface of the specimen was cathodically treated for 10 min with the potential on the potentiostat at −0.7 V (Ag/AgCl) in the solution. The specimen was then left for 10 min at its natural potential. Finally, the anodic polarization curve was measured by running the potentiostat from the natural potential to 1.1 V (Ag/AgCl) at 20 mV/min with 5 second intervals.

Each specimen was made of well-tempered 316L austenitic stainless steel. The chemical composition is shown in Table 1. They were square specimens with a thickness of 2.8 mm and side length of 35 mm as shown in Figure 3. The surfaces of the specimens were finished by several different surface finishes as shown in Table 2.

The effect of residual stress on current density was investigated after surface finishing; however, there are factors other than residual stress that vary with surface finishing that can affect the corrosion behavior. In order to investigate the effect of residual stress without altering the metallic structure of the specimen introduced by the surface finish, the residual stress was controlled by exposing the back of the specimen to a cavitating jet to generate curvature and introduce compressive stress at the front [21]. Figure 4 shows a schematic diagram of the cavitating jet apparatus. A high-speed water jet pressurized by a plunger pump is injected through a nozzle into a water filled tank, generating cavitation around the shear layer of the jet. The cavitating conditions were determined based on previous results [22]. The injection pressure of the cavitating jet and the pressure in the tank were set to 30 MPa and 0.42 MPa, respectively. In addition, the nozzle diameter and the standoff distance between the nozzle and the specimen were 2 mm and 85 mm, respectively. To vary the curvature generated by the cavitating jet, i.e., to vary the compressive residual stress, the specimens were exposed to the jet for various times (0, 10, 20, 40, 80 seconds), and the variation of the residual stress with curvature was evaluated by the $\sin^2\psi$ method.

Table 1: Chemical composition of the 316L austenitic stainless steel

Element	C	Si	Mn	P	S	Ni	Cr	Mo
Content (wt%)	0.014	0.630	0.970	0.030	0.004	12.030	17.450	2.050

Table 2: Variation of surface finish

Specimen	Surface finish
A	Electropolsih
B	Emery paper #800
C	Ceramic rubber polisher #240
D	Angle grinder A/WA36P

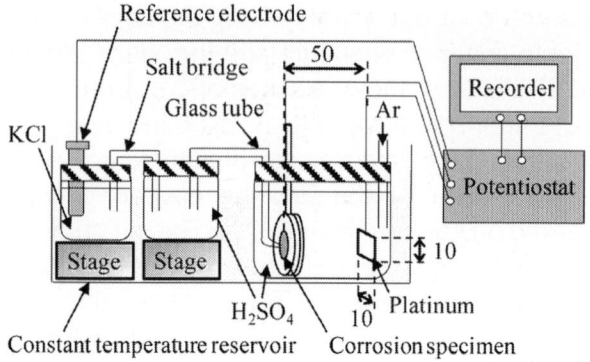

Figure 1: Apparatus for anodic polarization measurement.

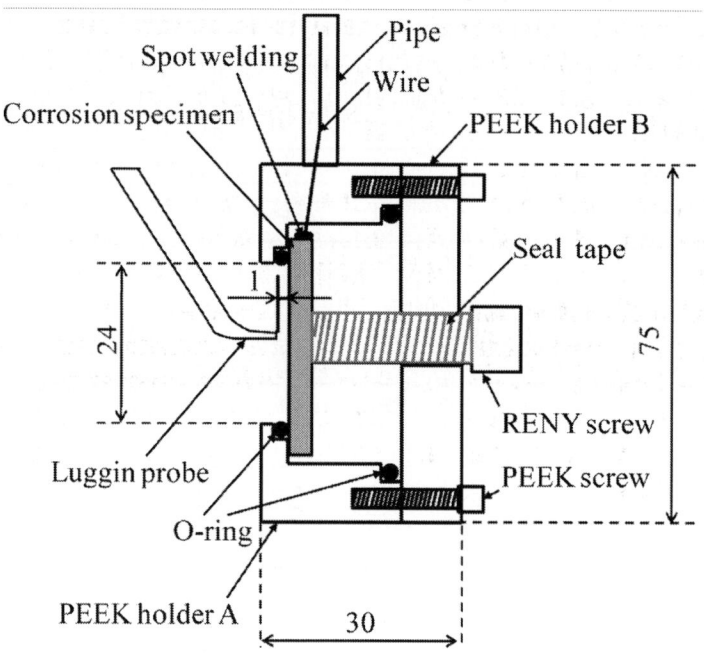

Figure 2: Specimen holder for electrochemical cell.

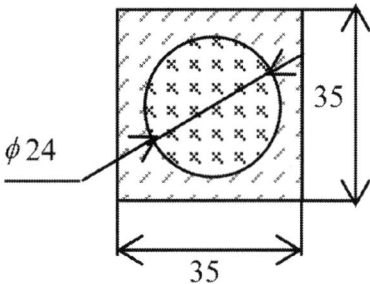

Figure 3: Geometry of the specimen.

The surface residual stress caused by the surface finish, $\sigma_{R'}$ was evaluated using an X-ray diffraction method employing an X-ray tube with a Cr target operated at 30 kV and 8 mA. X-rays from the Kβ peak were chosen. The angle of the solar slit was 1 degree and the slit width was 4 mm. The diffractive angle 2θ was varied from 143.5 to 153.5 degrees in steps of 0.2 degree, and the diffractive X-rays were counted for 3 sec. at each step using a scintillation counter at angles of ψ = 0, 22.8, 33.2, 42.2 and 50.8 degrees. The diffractive plane was the (3 1 1) plane of γ-Fe. The diffractive angle without strain $2\theta_0$ was 148.5 degrees, and the stress factor was −368.9 MPa/deg. The diffractive angle was determined by a half value width method, and the residual stress was calculated by a $\sin^2\psi$ method. In addition, in order to investigate the effect of the surface finish on the X-ray profile, it was obtained by employing a Cu target operated 40 kV and 40 mA. The diffractive plane was the (1 1 1) plane of γ-Fe. After measurements of the residual stress and X-ray profile, anodic polarization measurements were conducted using the above conditions.

RESULTS

Figure 5 shows the residual stresses at the surface generated by each surface finish. In Figure 5, the residual stresses, $\sigma_{R'}$ before surface finishing (electropolish) are 19 MPa, and after surface finishing they have several various values. In particular in the case of specimen D (Angle grinder), a large tensile residual stress was introduced, with σ_R = 315 MPa. The residual stress for specimens B (Emery paper #800), C (Rubber polisher #240) are 192 and 257 MPa in compression, in contrast to specimen D (Angle grinder). The difference in residual stress is due to the different surface finish, and this may generate different corrosion behavior. The following paragraph describes the differences in the anodic polarization curves obtained from the corrosion tests.

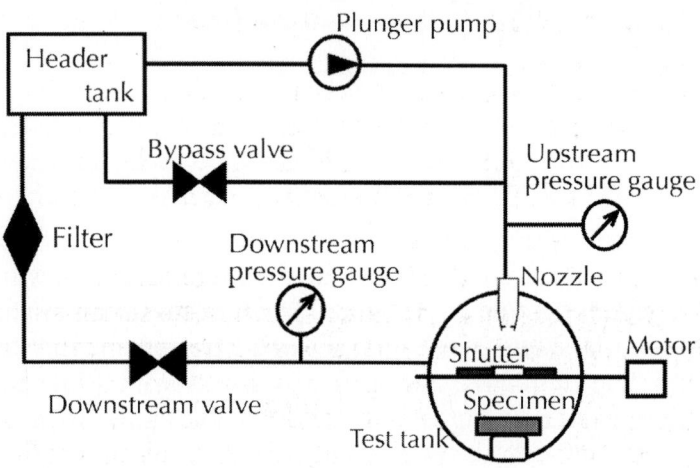

Figure 4: Apparatus for anodic polarization measurement.

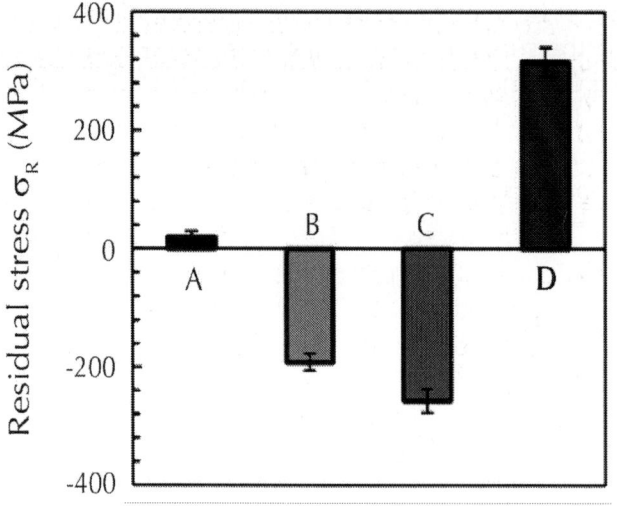

Figure 5: Residual stress introduced by each surface finish.

Figure 6 shows the current density as a function of electrode potential for each specimen. The current density was obtained by dividing the current by the test area. As shown in Figure 6, the current density varies depending on the surface finish. The surface finish affects the surface texture such as dislocation structure at surface produced by plastic deformation in the process of surface finish, grain size and residual stress at the metal surface. However, it is difficult to pick over the effect of those factors on the corrosion behavior, and the objective of this study is to investigate the effect of residual stress. At the same time, these other effects can be excluded by control of the residual stress only. We varied the residual stress without change of other factors, as described in the following paragraph.

Figure 7 shows the changes in residual stresses, $\Delta\sigma_R$, due to the application of a cavitating jet to the backs of the specimens as functions of exposure time, t. Despite the small differences, in each case, the residual stress becomes increasingly compressive, with increasing cavitating jet processing time. These increase up to a processing time of ~40 sec., and then saturate beyond that. Apart from specimen D (Angle grinder), $\Delta\sigma_R$ is about 80 MPa - 200 MPa. In the case of specimen D, $\Delta\sigma_R$ is almost 300 MPa. Figure 8 indicates that this method can be used to control the residual stress and can be applied to exclude factors other than residual stress which affect corrosion behavior.

Figure 8 shows the anodic polarization curves for each surface finish for various exposure times to the cavitating jet. Despite the small difference, the current density in the passive region decreases as a whole with increasing processing time for all specimens as shown in Figure 8. The current densities for each specimen at t =10 sec. and 20 sec. decrease rapidly from those at t = 0, and show the same tendencies as the variations in residual stresses shown in Figure 7. In order to clarify the effect of residual stress on the critical current density for passivation, i_{crit}, and the passive current density, i_{pass}, Figure 9 and Figure 10 show these as functions of residual stress for each specimen. As shown in Figure 9 and Figure 10, i_{crit} and i_{pass} decrease with the introduced compressive stress for all specimens. In the case of the specimen without a surface finish (specimen A; Electropolish), i_{crit} and i_{pass} are 27 $\mu A/cm^2$ and 9 $\mu A/cm^2$ at σ_R = 19 MPa (t = 0 s), and these parameters rapidly decrease with increasing compressive stress. Those parameters were saturated at 11 $\mu A/cm^2$ and 2 $\mu A/cm^2$ at σ_R= −202 MPa (t = 80 s). In all specimens, i_{crit} and i_{pass} shows same tendencies as the specimen A (Electropolish) and those parameters were decreased by more than 70% by applying a cavitating jet to the backs of the specimens to control the residual stress.

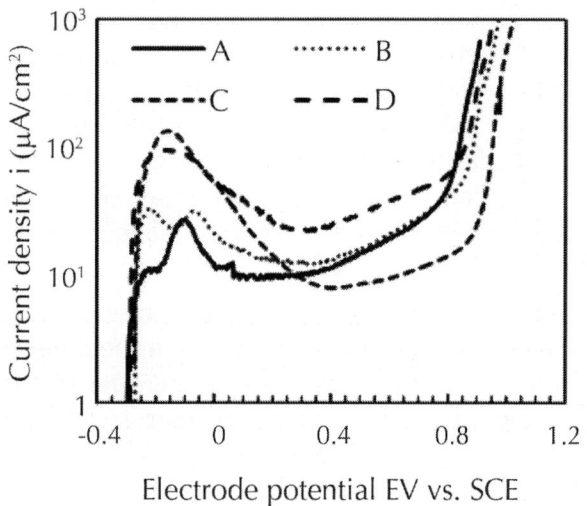

Figure 6: Anodic polarization curves for each surface finish.

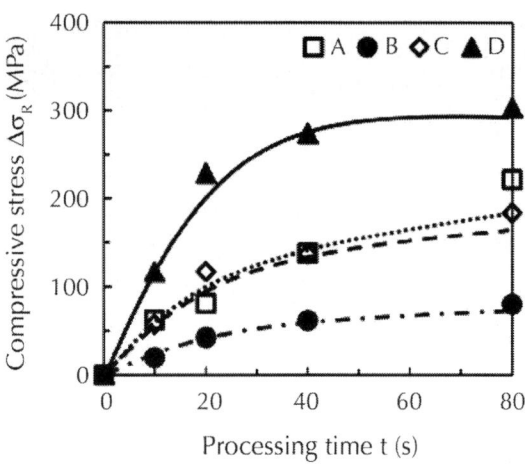

Figure 7: Change in residual stress with processing time of cavitating jet treatment at the back of the specimen.

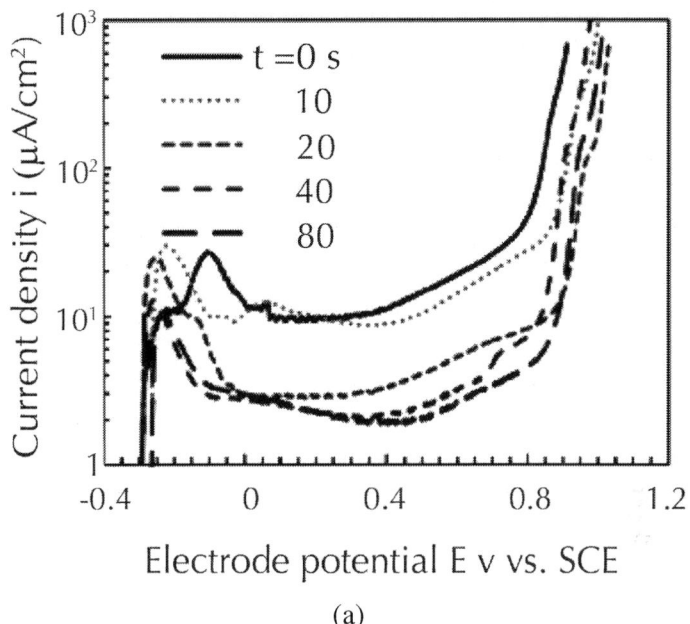

(a)

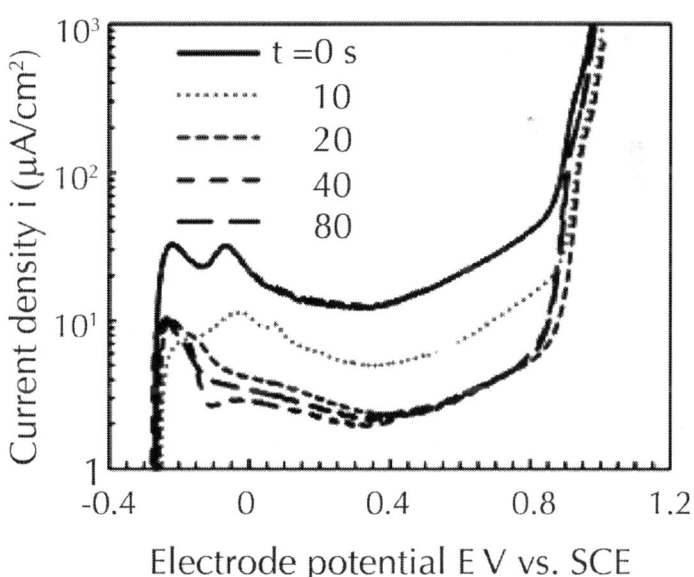

(b)

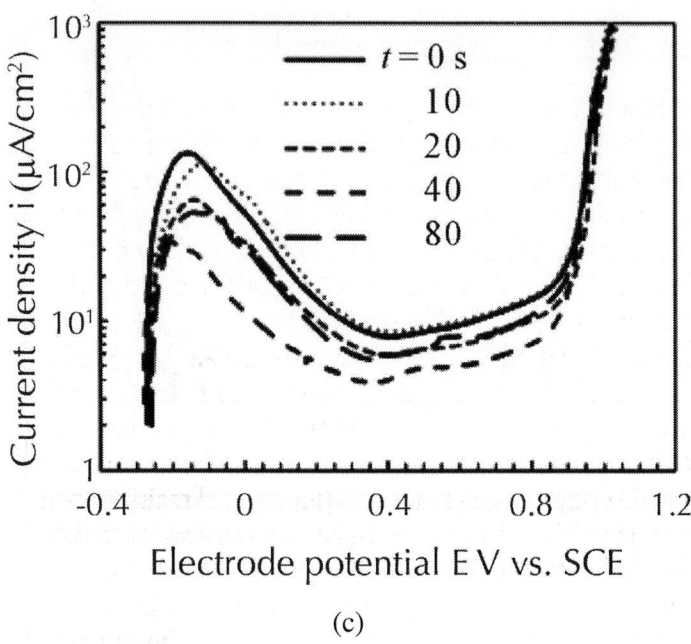

(c)

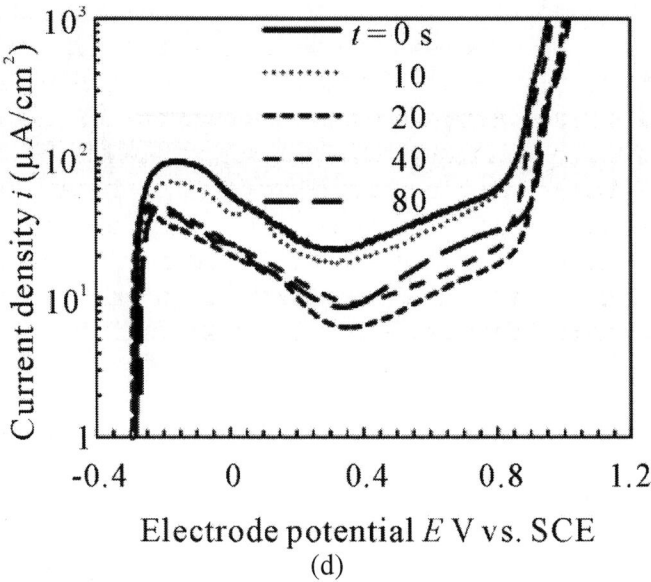

(d)

Figure 8: Change in anodic polarization curve after cavitating jet treatment at the back of the specimen.

DISCUSSION

The variation of the current density by the surface finish can be explained by the variation of surface activity. The surface finish affects surface conditions such as dislocation, grain size and residual stress etc. In particular, the residual stress is greatly affected by the condition of surface finish such as heat generated by the finishing process [23] [24]. Figure 11 shows an X-ray diffraction profile of the specimens finished by Electropolish (specimen A), Emery paper #800 (specimen B), Rubber polisher #240 (specimen C) and Angle grinder (specimen D). As shown in Figure 11, the surface finish largely affected X-ray diffraction profile. It is well known that the profile includes much information of the metal surface such as surface texture (dislocation structure, micro- strain in grain), grain size and residual stress etc. [25] [26]. The variation of current density depending on surface finish as shown in Figure 6 was attributed to those factors. Y. Wang reported that the surface activity has a large effect on the corrosion resistance of material and the reaction of dislocation on the surface affects the surface activity [27]. The variation of the anodic polarization curve due to the surface finish can be caused by the difference of surface activity, including dislocations, grain size and residual stress. In the view point of the grain size and metallic structure, the nanosize grains and strain induced martensite formed at the surface by shot-peening and ultrasonically peening methods, and they affected the anodic polarization curve, i.e., current density, of the 304 austenitic stainless steel [14] . Also, the high-density grain boundaries can promote the diffusion of chromium and it may strengthen the passive film [28]. It is quite likely that the variation of the current density depends on the surface texture which was changed by the surface finish.

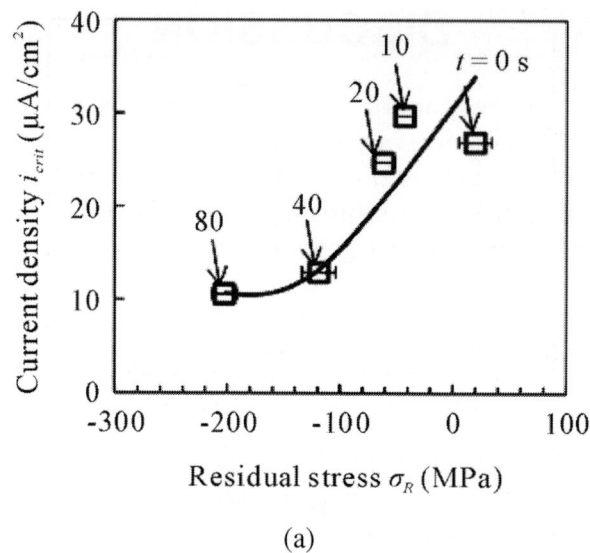

(a)

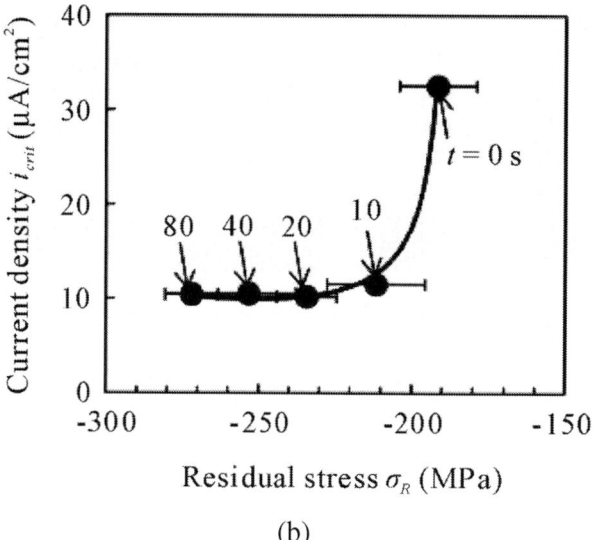

(b)

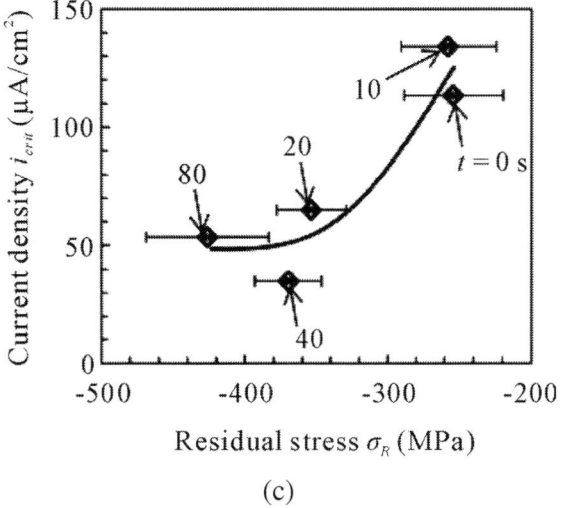

(c)

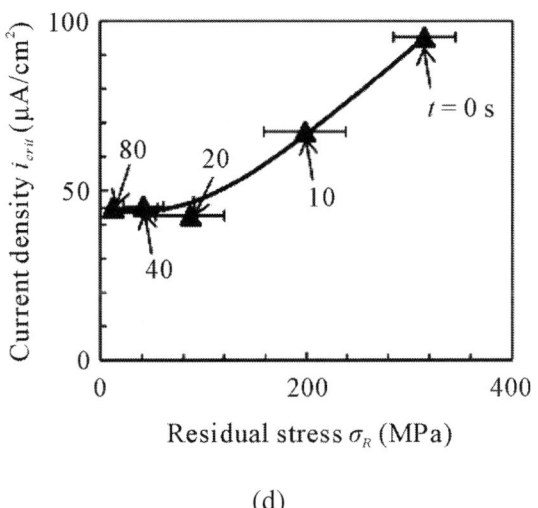

(d)

Figure 9: Critical current density for passivation decreasing with decreasing residual stress.

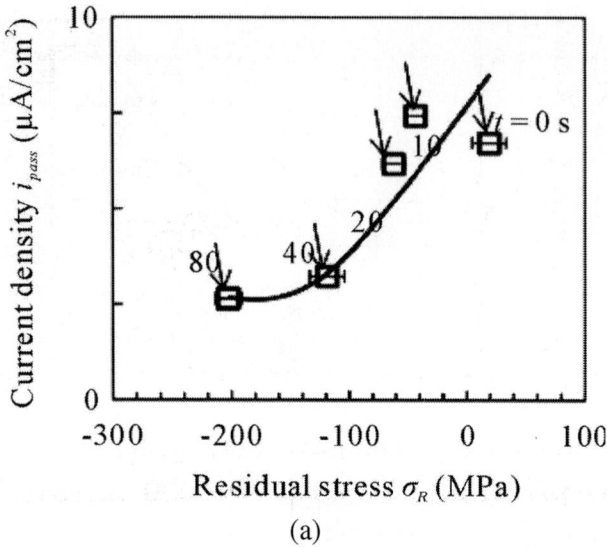

(a)

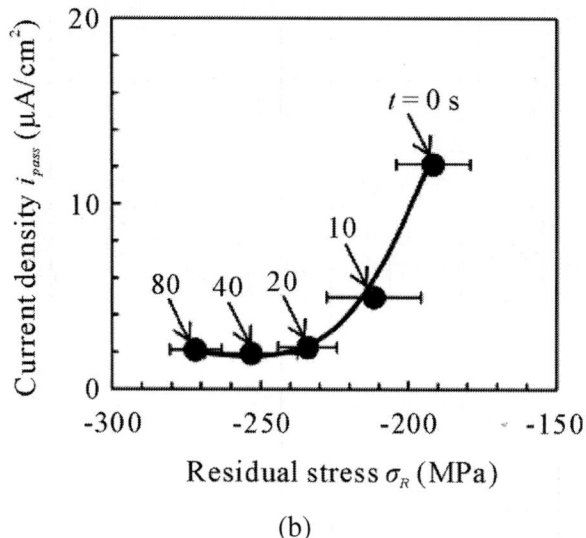

(b)

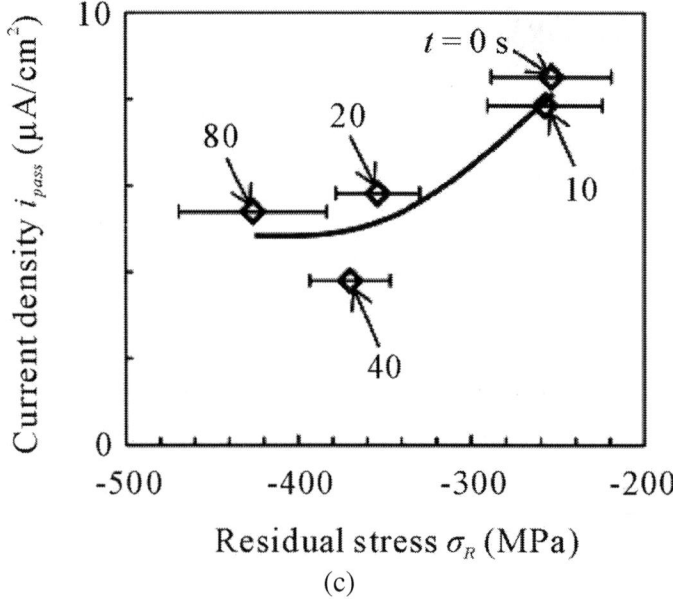

(c)

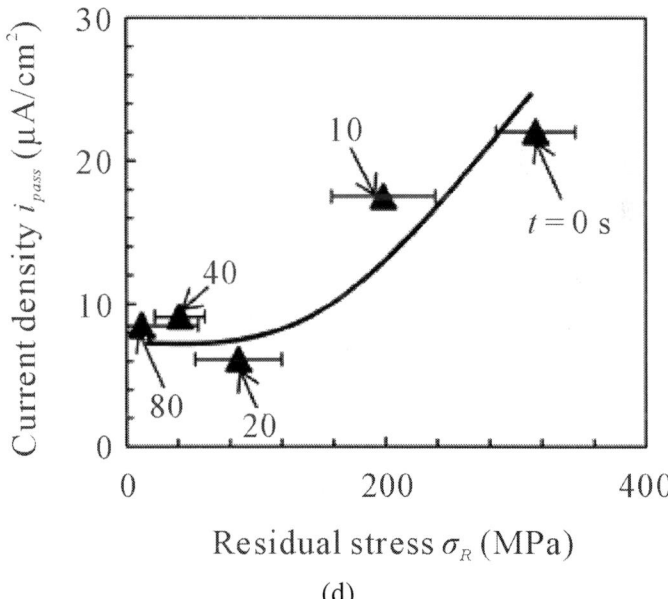

(d)

Figure 10: Passive current density for passivation decreasing with decreasing residual stress.

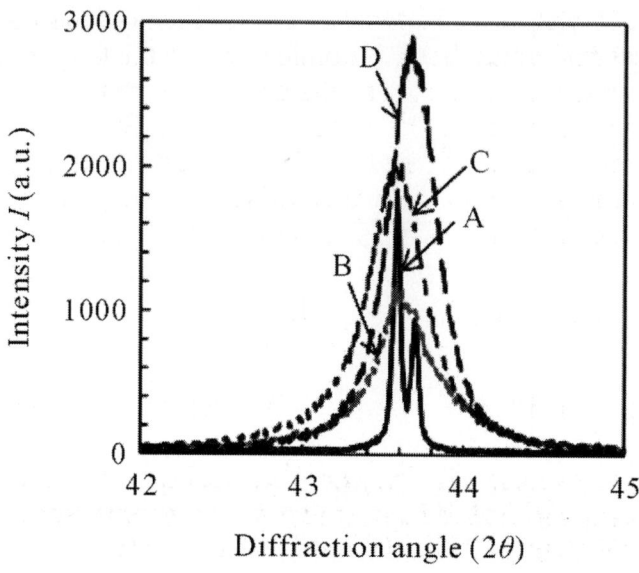

Figure 11: X-ray diffraction profile affected by surface finish.

Since the critical current density and the current density for passivation of all specimens were reduced by the introduction of compressive stress at surface without the variation of the surface texture, it can be considered with the exception of the factors other than the residual stress. The reduction of the current density by the introduction of compressive stress, as shown in Figure 8, Figure 9 and Figure 10, can be considered by the construction and growth of the passive film. The results indicate that the passive film could be constructed and kept by lower current density during the corrosion process compared to that of without the compressive stress. It was reported that the current density of aluminum alloy was reduced by the compressive stress and it can suppress rupture of the passive film contrary to the tensile stress [29]. Although there is a difference the construction behavior of the passive film between the stainless steel and the aluminum alloy, the role of compressive stress might be same as far as an auxiliary role for the construction and maintenance of the passive film goes. In addition, the current density of 316L stainless steel could be reduced by laser-peening and shot peening with introduction of large compressive residual stress, and the compressive residual stress assumes a role of the construction and growth of the passive film in austenitic stainless steel [13]. In viewpoint of an interatomic distance, the compressive residual stress narrows the interatomic distance contrary to the tensile residual stress and the thickening the chromium atoms at surface layer by the compressive residual stress can make it easy to construct, grow and keep the passive film. Further research based on the evaluation of the passive film varying with the residual stress would clarify the effect of residual stress on the corrosion behavior more precisely.

CONCLUSIONS

In this paper, in order to demonstrate the effect of residual stress on the corrosion behavior of austenitic stain- less steel, corrosion tests using an electrochemical method were conducted on specimens made of Japanese Industrial Standards SUS316L. The specimens were treated by several different surface finishes, and in order to exclude the effect of factors other than residual stress, such as surface texture, micro-strain and grain size, which can affect the corrosion behavior and thereby the generation of a passivation layer, the residual stress in each specimen was varied by generating curvature in it by applying a cavitating jet to the back of it. The conclusions obtained in the present study are summarized as follows.

- Surface finish caused the residual stress at surface and the anodic polarization curve was affected largely by the surface finish. In addition, the current density such as the critical current density for passivation and the passive current density was decreased rapidly with increasing the compressive stress. Depending on the type of surface finish, those current densities were decreased by more than 70% by introduction of compressive stress.

- The introduction of compressive residual stress makes producing the passivation film easier regardless of the surface condition varied by surface finish since the passive film can be produced and maintained at low current density. This might be the reason why the reduction of interatomic spacing due to the compressive stress at surface can facilitate the growth and maintenance of the passivation film. The compressive residual stress enhances not only the mechanical properties but also the corrosion resistance.

ACKNOWLEDGEMENTS

This work was partly supported by JSPS KAKENHI Grant number 24360040.

REFERENCES

1.   Boven, G.V., Chen, W. and Rogge, R. (2006) The Role of Residual Stress in Neutral pH Stress Corrosion Cracking of Pipeline Steels. Part I: Pitting and Cracking Occurrence. Acta Materialia, 55, 29-42.

2.   Mochizuki, M. (2007) Control of Welding Residual Stress for Ensuring Integrity against Fatigue and Stress-Corrosion Cracking. Nuclear Engineering and Design, 237, 107-123.http://dx.doi.org/10.1016/j.nucengdes.2006.05.006

3.   Soyama, H., Saito, K. and Saka, M. (2002) Improvement of Fatigue Strength of Aluminum Alloy by Cavitation Shot- Less Peening. Journal of Engineering Materials and Technology, Transactions of the ASME, 124, 135-139.http://dx.doi.org/10.1115/1.1447926

4.   Soyama, H., Shimizu, M., Hattori, Y. and Nagasawa, Y. (2008) Improving the Fatigue Strength of the Elements of a Steel Belt for CVT by Cavitation Shotless Peening. Journal of Materials Science, 43, 5028-5030. http://dx.doi.org/10.1007/s10853-008-2743-6

5.   Kobayashi, M., Matsui, T. and Murakami, Y. (1998) Mechanism of Creation of Compressive Residual Stress by Shot Peening. International Journal of Fatigue, 20, 351-357.http://dx.doi.org/10.1016/S0142-1123(98)00002-4

6.   Soyama, H., Kikuchi, T., Nishikawa, M. and Takakuwa, O. (2010) Introduction of Compressive Residual Stress into Stainless Steel by Employing a Cavitating Jet in Air. Surface and Coatings Technology, 205, 3167-3174.http://dx.doi.org/10.1016/j.surfcoat.2010.11.031

7.   Sano, Y., Akita, K., Masaki, K., Ochi, Y., Altenberger, I. and Scholtes, B. (2006) Laser Peening without Coating as a Surface Enhancement Technology. Journal of Laser Micro/Nanoengineering, 1, 161-166. http://dx.doi.org/10.2961/jlmn.2006.03.0002

8.   Sedriks, A.J. (1996) Corrosion of Stainless Steels. 2nd Edition, Wiley-Interscience, New York.

9.   Begujm, Z., Poonguzhali, A., Basu, R., Sudha, C., Shaikh, H., Rao, R.V.S., Patil, A. and Dayal, R.K. (2011) Studies of the Tensile and Corrosion Fatigue Behaviour of Austenitic Stainless Steels. Corrosion Science, 53, 1424-1432.http://dx.doi.org/10.1016/j.corsci.2011.01.003

10.   Davoodi, A., Pakshir, M., Babaiee, M. and Ebrahimi, G.R., (2011) A Comparative H2S Corrosion Study of 304L and 316L Stainless Steels in Acidic Media. Corrosion Science, 53, 399-408. http://dx.doi.org/10.1016/j.corsci.2010.09.050

11.   Vignal, V., Zhang, H., Delrue, O., Heintz, O., Popa, I. and Peultier, J. (2011) Influence of Long-Term Ageing in Solution Containing Chloride. Corrosion Science, 53, 894-903.http://dx.doi.org/10.1016/j.corsci.2010.11.011

12.   Garcia, C., Tiedra, M.P., Blanco, Y., Martin, O. and Martin, F. (2008) Intergranular Corrosion of Welded Joints of Austenitic Stainless Steels Studied by Using an Electrochemical Minicell. Corrosion Science, 50, 2390-2397.http://dx.doi.org/10.1016/j.corsci.2008.06.016

13.   Peyre, P., Scherpereel, X., Berthe, L., Carboni, C., Fabbro, R., Beranger, G. and Lemaitre, C. (2000) Surface Modifications Induced in 316L Steel by Laser Peening and Shot Peening. Influence on Pitting Corrosion Resistance. Materials Science and Engineering A, 280, 294-302. http://dx.doi.org/10.1016/S0921-5093(99)00698-X

14.   Lee, H., Kim, D., Jung, J., Pyoun, Y. and Shin, K. (2009) Influence of Peening on the Corrosion Properties of AISI 304 Stainless Steel. Corrosion Science, 51, 2826-2830.http://dx.doi.org/10.1016/j.corsci.2009.08.008

15.   Azer, V., Hashemi, B. and Yazdi, M.R. (2010) The Effect of Shot Peening on Fatigue and Corrosion Behaviour of 316L Stainless Steel in Ringer's Solution. Surface and Coatings Technology, 204, 3546-3551. http://dx.doi.org/10.1016/j.surfcoat.2010.04.015

16.   Bahadur, A., Kumar, B.R. and Chowdhury, S.G. (2004) Evaluation of Changes in X-Ray Elastic Constants and Residual Stress as a Function of Cold Rolling of Austenitic Steels. Materials Science and Technology, 20, 387-392.http://dx.doi.

org/10.1179/026708304225012170

17. Kumar, B.R., Mahato, B. and Singh, R., Kumar, B.R., Mahato, B. and Singh, R. (2007) Influence of Cold-Worked Structure on Electrochemical Properties of Austenitic Stainless Steels. Metallurgical and Materials Transactions A, 38, 2085-2094. http://dx.doi.org/10.1007/s11661-007-9224-4

18. Peguet, L., Malki, B. and Baroux, B. (2009) Effect of Austenite Stability on the Pitting Corrosion Resistance of Cold Worked Stainless Steels. Corrosion Science, 51, 493-498.http://dx.doi.org/10.1016/j.corsci.2008.12.026

19. Ghanem, F., Braham, C., Fitzpatrick, M.E. and Sidhom, H. (2002) Effect of Near-Surface Residual Stress and Micro-Structure Modification from Machining on the Fatigue Endurance of a Tool Steel. Journal of Materials Engineering and Performance, 11, 631-639. http://dx.doi.org/10.1361/105994902770343629

20. Turnbull, A., Mingard, K., Lord, J.D., Roebuck, B., Tice, D.R., Mottershead, K.J., Fairweather, N.D. and Bradbury, A.K. (2011) Sensitivity of Stress Corrosion Cracking of Stainless Steel to Surface Machining and Grinding Procedure. Corrosion Science, 53, 3398-3415. http://dx.doi.org/10.1016/j.corsci.2011.06.020

21. Takakuwa, O., Kawaragi, Y. and Soyama, H. (2013) Estimation of the Yield Stress of Stainless Steel from the Vickers Hardness Taking Account of the Residual Stress. Journal of Surface Engineered Materials and Advanced Technology, 3, 262-268.http://dx.doi.org/10.4236/jsemat.2013.34035

22. Soyama, H. (2011) Enhancing the Aggressive Intensity of a Cavitating Jet by Means of the Nozzle Outlet Geometry. Journal of Fluids Engineering, Transactions of the ASME, 133, Article ID: 101301.

23. Outeiro, J.C., Pina, J.C., Saoubi, R.M., Pusavec, F. and Jawahir, I.S. (2006) Analysis of Residual Stresses Induced by Dry Turning of Difficult-to-Machine Materials. International Journal of Machine Tools and Manufacture, 46, 1786- 1794.http://dx.doi.org/10.1016/j.ijmachtools.2005.11.013

24. Lazoglu, I., Ulutan, D., Alaca, B.E., Engin, S. and Kaftanoglu, B. (2008) An Enhanced Analytical Model for Residual Stress Prediction in Machining. Manufacturing Technology, 57, 81-84.

25. Ungar, T. and Borbely, A. (1996) The Effect of Dislocation Contrast on X-Ray Line Broadening: A New Approach to Line Profile Analysis. Applied Physics Letters, 69, 3173-3175. http://dx.doi.org/10.1063/1.117951

26. Barabash, R. (2001) X-Ray and Neutron Scattering by Different Dislocation Ensembles. Materials Science and Engineering A, 309-310, 49-54. http://dx.doi.org/10.1016/S0921-5093(00)01663-4

27. Wang, Y., Zhao, W., Ai, H., Zhou, X. and Zhang, T. (2011) Effects of Strain on the Corrosion Behaviour of X80 Steel. Corrosion Science, 53, 2761-2766.http://dx.doi.org/10.1016/j.corsci.2011.05.011

28. Wang, X.Y. and Li, D.Y. (2002) Mechanical and Electrochemical Behavior of Nanocrystalline Surface of 304 Stainless Steel. Electrochimica Acta, 47, 3939-3947.http://dx.doi.org/10.1016/S0013-4686(02)00365-1

29. Liu, X. and Frankel, G.S. (2006) Effects of Compressive Stress on Localized Corrosion in AA2024-T3. Corrosion Science, 48, 3309-3329.http://dx.doi.org/10.1016/j.corsci.2005.12.003

Electrochemical Investigation of Corrosion on AISI 316 Stainless Steel and AISI 1010 Carbon Steel: Study of the Behaviour of Imidazole and Benzimidazole as Corrosion Inhibitors

Roberta R. Moreira, Thiago F. Soares,
and Josimar Ribeiro

Department of Chemistry, Federal University of Espírito Santo, Vitória,
Brazil

ABSTRACT

An electrochemical investigation of the corrosion on AISI 316 austenitic stainless steel and AISI 1010 carbon steel in sodium chloride solution (3.0 wt.%) was performed in the absence and presence of imidazole and benzimidazole corrosion inhibitors. The results showed that at any inhibitor concentration (25 ppm to 1000 ppm), there was an increase in the polarisation resistance of both steels. The highest efficiency of corrosion inhibition was obtained using imidazole at a concentration of 50 ppm for both steels, with values of 96% for the AISI 316 stainless steel and 73% for the AISI 1010 carbon steel.

INTRODUCTION

Corrosion as a destructive phenomenon incurs a high cost to the economy and society. The costs have been estimated to be as high as 5% of GDP in many developed countries [1] [2] . All metallic material can undergo different types of degradation processes when exposed to different environmental conditions, either by natural factors or by industrial or other human activities [3] . Corrosion can affect critical areas or regions of a material and thereby modify its true functionality [4] . Additionally, the lifetime of metallic equipment tends to decrease further in more aggressive environments [1] .

In coastal regions, the corrosion of steel infrastructure can have considerable negative effects, and due to the high costs of maintenance and corrosion protection measures, there is interest in applying increasingly resistant steels [5] . According to Tsutsumi et al. [6] , sea water is an aggressive environment that affects almost all structural materials. Present in the composition of sea water are salts such as $NaCl$ and $MgCl_2$, which are primary agents responsible for localised corrosion. The chloride ion is the most common in marine environments and the main cause of the breakdown of the passive film on steels [7] [8] . The presence or absence of this protective film on the surface of metals is responsible for controlling the behaviour of corrosion because of its action as a protective barrier against corrosion attacks [4] [9] .

Several metals suffer pitting corrosion when exposed to solutions with high chloride content [10] . Pitting corrosion is characterised by a process of localised metal dissolution that affects the integrity of the metal and, in some cases, leads to catastrophic failure. Although this form of corrosion occurs only in small areas of the metal surface, it remains a major concern due to its fast attack and penetration inside the metal [11] [12] .

The increased use of stainless steel has contributed substantially to the economic development of many countries. For decades, it has been used structurally in architectural elements such as beams and arches in bridges [13] .

In the case of stainless steel in contact with seawater, the corrosion rate and pitting potential depend on the levels of Cr and Ni present in the steel. The presence of Co, Mo and N has a significant influence on the corrosion resistance [14] . Austenitic stainless steels, whose main alloying elements are chromium and nickel, are considered to have high corrosion resistance in most environments compared with ferritic and martensitic steels [15] . The AISI 316 stainless steel, specifically, is perhaps one of the most used steels, after carbon steel, serving a structural material in industrial plants in marine environments, including desalination plants [14] . Carbon steel is also widely used in systems that operate in land and marine environments [16] and is found in various applications in our everyday life due to its relatively low cost [17] .

To preserve the integrity of metallic materials, some measures to inhibit or prevent corrosion in many aggressive media are cited in the literature [4] [18] . Organic inhibitors that exhibit one or more polar functions (such as N, O and S) and heterocyclic compounds with polar groups and ϖ electrons have been quite effective in corrosion protection [19] . This inhibition efficiency is usually attributed to the specific interactions that occur between functional groups and heteroatoms with the metal surface due to their lone pair electrons [20] and the influence on the change of corrosion potential [20] [21] . Certain organic compounds, such as thiophene derivatives, were studied as corrosion inhibitors for AISI 316 stainless steel in acid media with chloride ions and resulted in an efficiency of approximately 97% [22] . According to studies conducted by Farahani and Goudarzi [23] , a benzothiazole derivative was also efficient in inhibiting the corrosion of AISI 316 stainless steel in an acid medium.

With respect to AISI 1010 carbon steel, a compound commercially known as Dodigen, which is a mixture of amines and amides, was studied as a corrosion inhibitor by Granero et al. [24] . The authors found that Dodigen had an efficiency ranging from 74% to 98%, depending on the concentration used. Other investigations using computer simulations of new imidazole derivatives showed that these compounds possess the ability to inhibit the corrosion process on AISI 1010 carbon steel, which showed 62.8% inhibition efficiency [25] .

Other organic inhibitors such as benzimidazole and its derivatives have been reported in the literature as effective corrosion inhibitors with structural features that allow adsorption to the metal surface [26] [27] . Benzimidazole, an imidazole derivative, is an organic compound that has aromaticity and nitrogen heteroatoms with lone pair electrons, which facilitate the adsorption to the metal [27] - [29] . Studies have shown that it provides a high corrosion inhibition ability to copper in acidic, neutral and alkaline media [26] and to carbon steel in acidic medium [30] .

Imidazole and its derivatives have been used as corrosion inhibitors for various metals and alloys [30] - [33] . The advantages of imidazole, including that it does not have a detrimental effect on the environment and its convenience and low cost, have made the use of these compounds very attractive [32] - [37] . Imidazole is a planar organic five-membered aromatic heterocycle that has two nitrogen atoms [31] . The protective mechanism of this molecule to the surface of the metallic material can occur via the adsorption of electron pairs isolated from the pyrrole nitrogen or parallel to the surface [36] of the imidazole ring.

This paper investigated the corrosion of AISI 316 austenitic stainless steel and AISI 1010 carbon steel in solutions containing 3.0 wt% chloride ions and the efficiency of corrosion inhibition of benzimidazole and imidazole inhibitors in different concentrations. The morphological and compositional analyses were performed by Laser-induced plasma emission spectroscopy (LIPS), X-ray diffraction (XRD), scanning electron microscopy (SEM) and energy dispersive X-ray (EDX), and the electrochemical study was performed by potentiodynamic polarisation.

EXPERIMENTAL SECTION

Sample Preparation

The samples used in this study were AISI 316 austenitic stainless steel, provided by Tecinox industry, and AISI 1010 carbon steel, provided by the Brazil Steel RDG; both companies are located in Espírito Santo, Brazil. The samples were prepared in two different ways. For morphological analysis, the samples were prepared in a square shape, with dimensions of 1.0 cm × 1.0 cm. For the electrochemical investigations, the samples were L-shaped, with dimensions of 3.0 cm × 1.0 cm. Each sample was prepared to leave an area of 1.0 cm^2 to test, which was isolated with the aid of an epoxy resin (Araldite®).

Before performing each analysis, the samples were subjected to cleaning and polishing of the surface using sandpaper of several different particle sizes in water, respectively, in the following order: 220, 320, 400, 600 and 1200 mesh. The polishing was finished with 0.3 μm alumina to obtain a uniform surface. At each change of sanding, the sample was rotated 90°. The specimens were then washed with distilled water, degreased with acetone and dried with a hot air blast. Before and after every analysis, the samples were stored in argon to prevent contamination by contact with the external environment.

Physico-Chemical Characterisation

Laser-induced plasma emission spectroscopy using a Foundry-Master Pro spectrometer of the Shimadzu Corporation equipped with an Echelle optics and Kodak KAF 1001 ICCD detector allowed us to estimate the chemical composition of the alloys.

The XRD analyses were carried out on a Bruker model D8 Discover diffractometer operating with Cu K radiation (= 1.5406 Å) generated at 40 kV and 40 mA; 2 was scanned from 20° to 100° at a scan rate of 2° min^{-1}. The EVA V3.1 software was used to process the XRD data.

Metallographic analysis was performed to reveal the microstructure of steels. A chemical attack was performed on the sample of AISI 316 austenitic stainless steel with 1:3 HNO_3:HCl solution for 20 seconds to reveal the structure of the austenite grains and the presence of carbides. In the case of AISI 1010 carbon steel, 2% Nital (98% ethyl alcohol and 2% nitric acid) was used as the attack solution for 15 s to reveal the grain boundaries of ferrite. Microstructural characterisation was performed using a Nikon Eclipse 200 MA optical microscope at 1000× magnification with polarised light. A Shimadzu SS550 scanning electron microscope coupled to an SEDX model analyser was used to investigate the corrosion morphology. The EDX results were obtained from the matrix interference, atomic number, absorbance, and fluorescence (ZAF correction) data. Between the electrochemical and morphological analyses, all of the samples were stored in argon atmosphere to avoid any outside contamination.

Electrochemical Characterisation

Electrochemical measurements were performed using a 100 mL capacity electrochemical cell with three electrodes. The reference electrode was Ag/AgCl$_{(KCl,\ 3mol/L)}$, the counter electrode was carbon with dimensions of 2.1 cm × 0.5 cm × 1.8 cm, and the working electrode was made of AISI 316 austenitic stainless steel or AISI 1010 carbon steel. For electrochemical tests, a potentiostat/galvanostat AUTOLAB PGSTAT model 302N and GPES software were used. The corrosion tests were performed in duplicate using reagents to prepare solutions and analytical grade deionised water with a conductivity of 23.07 mS·cm^{-1} at 25°C from a deioniser trademark of Union analysis. The temperature was monitored and remained at 24°C ± 1.0°C. The corrosive media used in the tests were composed of sodium chloride at 3.0 wt% using PA-Impex reagent ACS in the presence or absence of benzimidazole (Sigma-Aldrich, 98%) and/or imidazole (Sigma-Aldrich, 98%). Imidazole and benzimidazole were employed as corrosion inhibitors at different concentrations: 25, 50, 100, 500 and 1000 ppm. For Tafel polarisation measurements, the potential was varied by ±250 mV from the open circuit potential (E_{ocp}) at a scan rate of 0.5 mV·s^{-1}. E_{ocp} was obtained by measuring electrochemical noise and was measured until the system came into balance for a period of 4000 seconds.

RESULTS AND DISCUSSION

Laser-Induced Plasma Emission Spectroscopy (LIPS)

Table 1 shows the results for the chemistry of the steels obtained via LIPS. The results obtained for the studied steels are within the composition range of AISI 316 [15] and AISI 1010 [38] as described in the literature.

X-Ray Diffraction

The XRD patterns of the analysed samples of AISI 316 stainless steel and AISI 1010 carbon steel are shown in Figure 1. The possible crystallographic phases present in each sample of steel and the orientation of these crystals can be observed. The XRD pattern of AISI 316 stainless steel shows diffraction peaks at 43.59°, 74.65° and 90.63°, relative to iron-nickel phase (FeNi) with reflection planes (111), (220) and (311), respectively, as the PDF-[01-077-7974]. A peak relative to chromium (Cr) phase with reflection plane (200) in accordance with the PDF-[01-088-2323] was observed at 50.78°. Both phases present have a face-centred cubic structure.

The XRD pattern of AISI 1010 carbon steel contains peaks that can be attributed to iron (Fe) present in a single phase composed of ferrite whose 2 angles are 44.71°, 65.08° and 82.41° with reflection planes (110), (200) and (211), respectively, as the PDF-[06-0696]. The observed phase structure is body-centred cubic.

Physico-Chemical Investigation

The metallography images of the studied steels are presented in Figure 2 and show the grain boundaries of the phases present in the steel and the presence of carbides. The image obtained for the AISI 316 steel (Figure 2(a)) shows that the chemical attack revealed the grain boundaries of the austenite phase grains of precipitated carbides and the presence of twinned boundaries, which is caused by the displacement of atoms in the lattice due to stress or heat treatment.

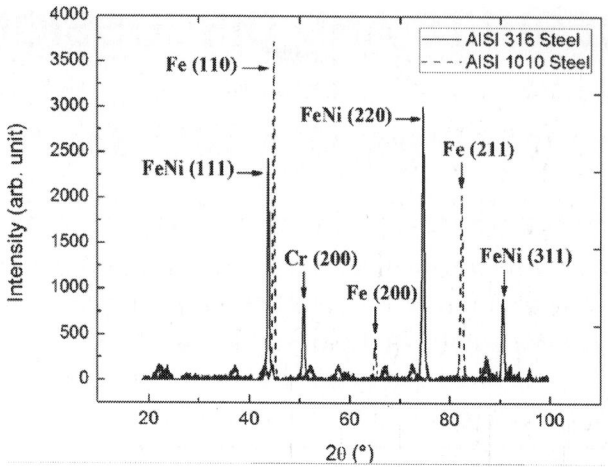

Figure 1. XRD pattern of AISI 316 stainless steel and AISI 1010 carbon steel.

Table 1. Chemical composition of AISI 316 austenitic stainless steel and AISI 1010 carbon steel

AISI	Chemical Composition(Wt%)							
	C	Mn	Si	Cr	Mo	P	S	Ni
316	0.0501	1.34	0.0507	16.3	2.01	0.0278	0.0019	10.5
1010	0.0954	0.351	0.0083	0.0238	0.0023	0.0150	0.0084	0.0074

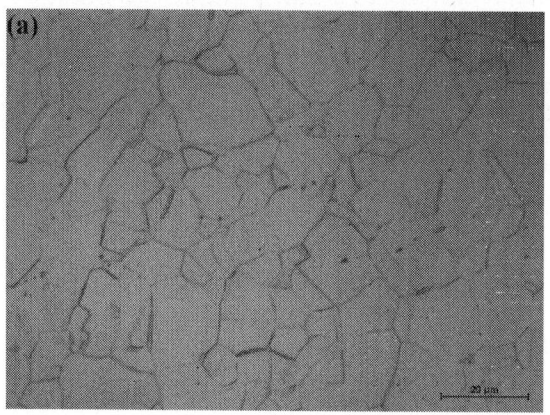

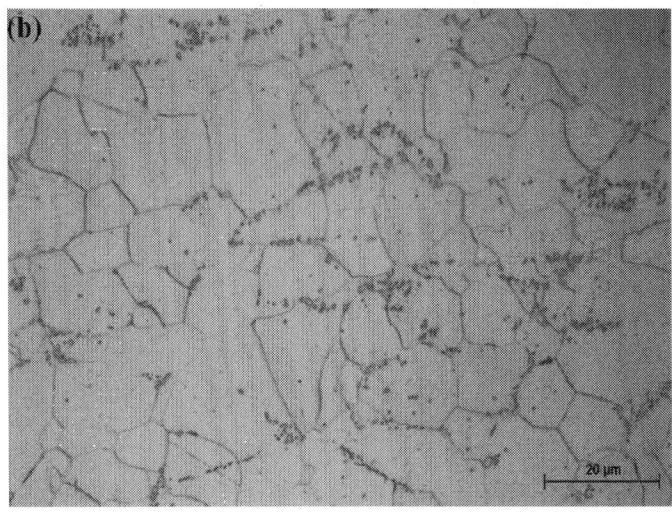

Figure 2. Metallographic images at 1000× magnification of the two grades of steel before the attack of chloride ions in solution. (a) AISI 316 stainless steel and (b) AISI 1010 carbon steel.

Similarly, the metallographic image of AISI 1010 steel (Figure 2(b)) revealed the grain boundaries present, which are composed of ferrite. Higher than the amounts of AISI 316 were also revealed that grain carbide steel, including regions of the grain boundary, which can be observed in Figure 2(b). Such regions, including the steel grain boundaries, deformation and non-metallic inclusions, have been found in the literature to be preferred sites for the occurrence of corrosion, particularly pitting corrosion [16] [39] .

The scanning electron microscopy and EDX spectra of the steel before the etching process are shown in Figure 3. Nonmetallic inclusions are present in both of the steel samples studied. These inclusions are more prone to the start of an anodic attack in a corrosive environment, and these areas become the sites of nucleation of pitting [16] [39] . The difference in composition between regions of inclusions and the matrix of the steel is responsible for generating the potential difference in these two areas, which initiates the corrosion process.

The EDX analysis of AISI 316 (Figure 3(b)) stainless steel shows the elemental composition of the sample found in the possible inclusion of steel and the elemental composition of the matrix of the steel. From this analysis, it is observed that this inclusion in the AISI 316 steel is composed mostly of carbon and oxygen. The EDX analysis also shows peaks related to silicon, manganese, sulphur and chromium in lower content than the matrix, showing that it can accommodate Si and Cr oxides, as well as the presence of carbides and manganese sulphide. The presence of grains of carbides precipitated in the steel is in accordance with the analytical results of the inclusions performed using the metallography technique, as mentioned above.

Elements such as oxygen, sulphur, nitrogen, phosphorus, chlorine, sodium and potassium may exist in steel primarily as a result of the steel-making process and are responsible for generating negative effects on the properties of steels, causing brittleness and lower corrosion resistance. To reduce the amount of oxygen and sulphur in Al-Si alloy, for example, elements are added in the metal degreasing process because of the strong reactivity of these elements to oxygen forming their oxides [40] .

Figure 4 shows EDX spectra and micrographs of the steels after the corrosion process in a solution containing 3.0 wt% chloride ions. The SEM analysis of the surface of the steels after suffering corrosive attack by chloride ions allowed the identification of the type of corrosion that occurred on the surface of each of the steels. After analysing the image in Figure 4(a), which shows the surface of the AISI 316 steel after attack by chloride ions, the type of corrosion was identified as the pitting type, which is characterised by a larger corrosion depth than diameter and occurs in small, localised regions of the metal surface. However, the image that shows the AISI 1010 steel surface after attack by chloride ions (Figure 4(b)) reveals that there has been widespread and uniform corrosion on the steel surface because the corrosion is present across the extent of the surface and because the corrosion product exhibits porosity.

The corrosive attack of chloride ions to the AISI 316 steel was expected because pitting corrosion occurs in passive metals and steel in regions where there are flaws in the protective film, as observed in studies by Burstein and Pistorius [10] , Kondo [11] and Turnbull et al. [12] , and in deformed regions and non-metallic inclusions [16] [39] . This explanation is supported by SEM surface analysis and by EDX analysis of the pitting, which shows the presence of elements related to non-metallic inclusions (Figure 4(c)) and that this steel is susceptible to such corrosion characteristics. Conversely, because AISI 1010 steel is not passivated, the entire sur- face is exposed to the corrosive medium and is thus susceptible to widespread corrosion, instead of localised pitting corrosion, as observed in the surface analysis. Additionally, the EDX results (Figure 4(d)) contain no evidence of inclusions containing Al, Si or S, thereby reducing the likelihood of localised pitting corrosion.

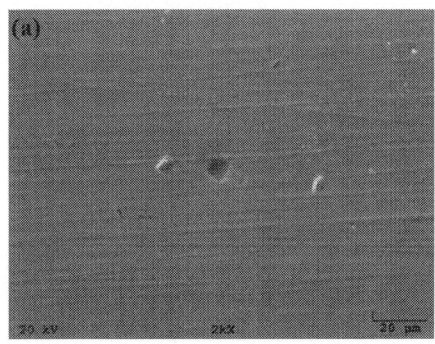

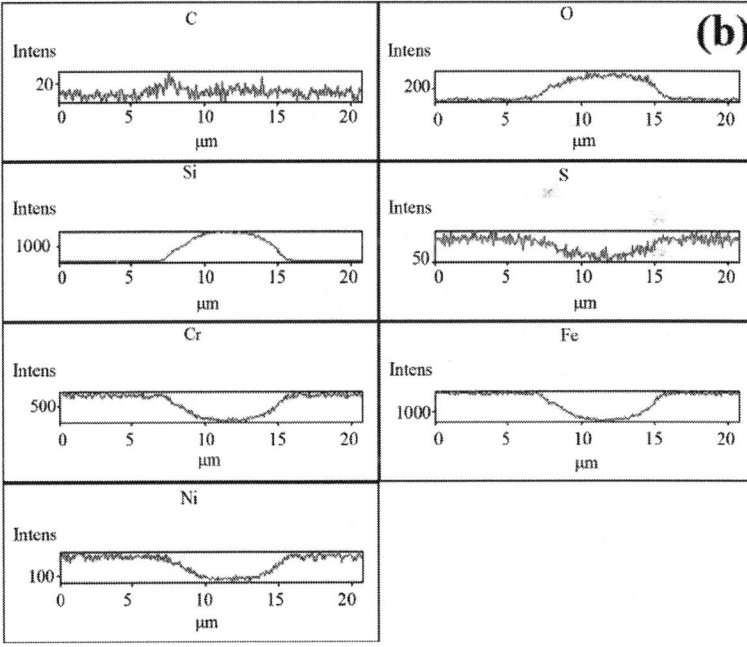

Figure 3. SEM images (a) and EDX analyses (b) of sample of AISI 316 steel at 2000× magnification before the corrosion process.

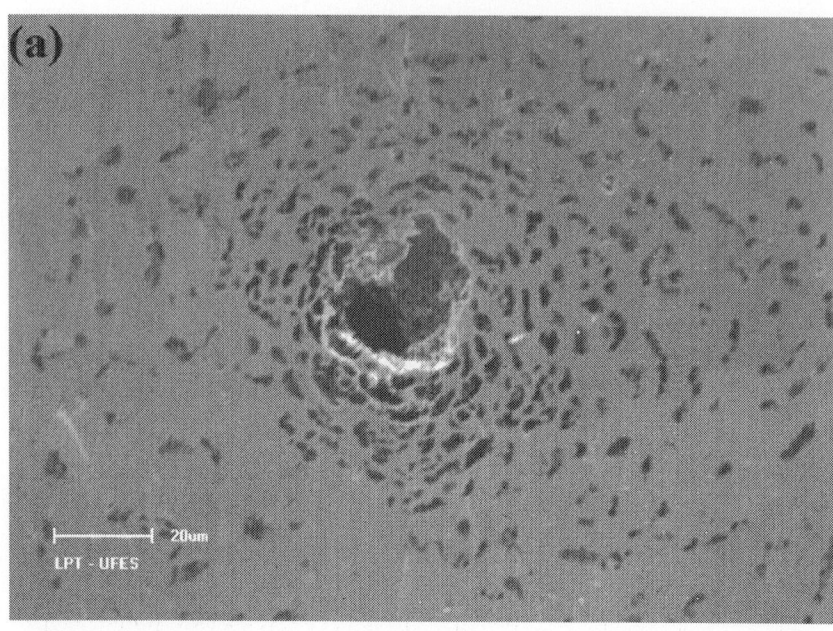

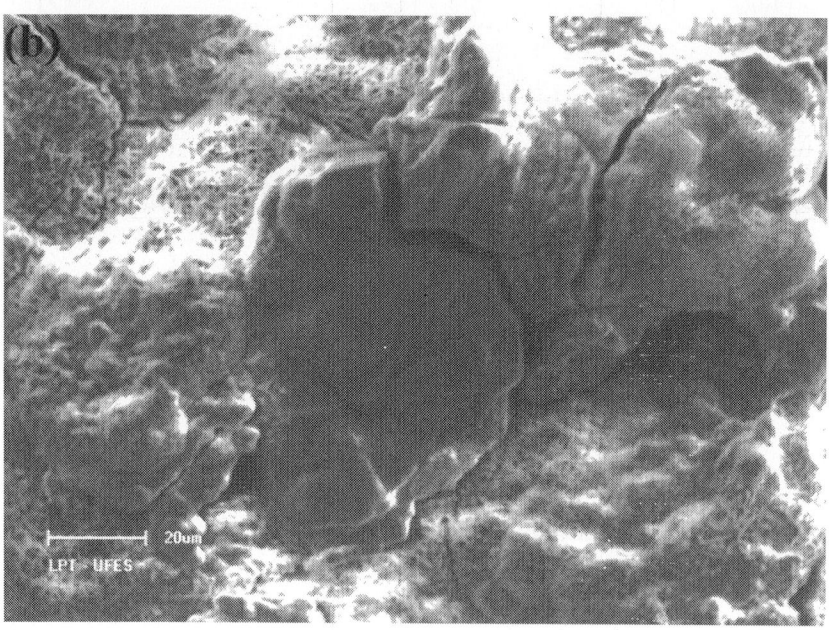

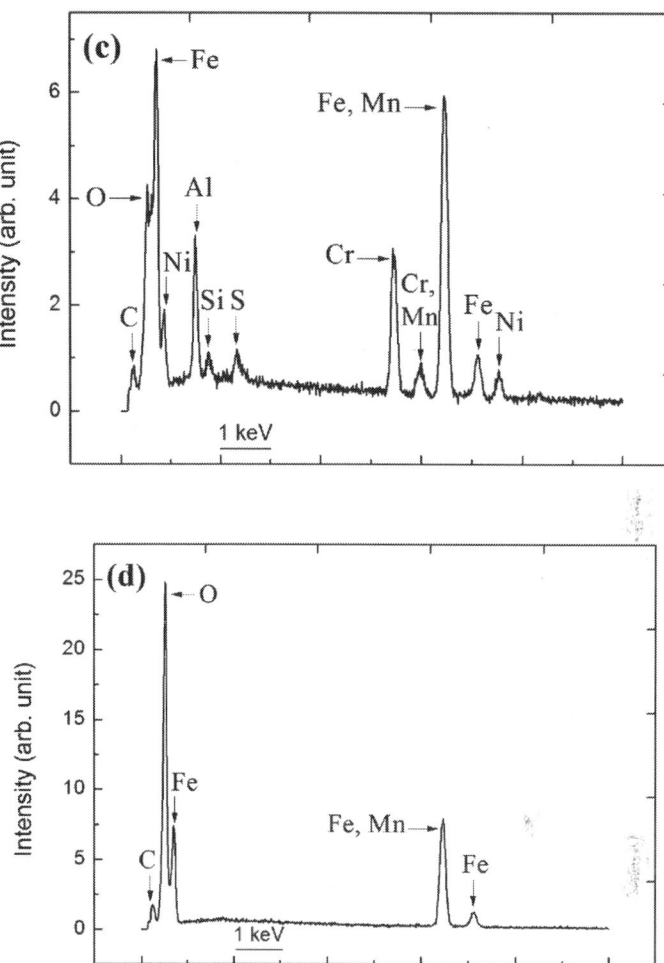

Figure 4. SEM images and EDX analyses of steel samples at 600× magnification after the corrosion process in the presence of sodium chloride at 3.0 wt.% solution. (a) (c) AISI 316 stainless steel; (b) (d) AISI 1010 carbon steel.

Electrochemical Investigations

The electrochemical behaviour of AISI 316 stainless steels and AISI 1010 carbon steel in the presence of chloride ions was investigated using Tafel plots, as shown in Figure 5. From Tafel analysis of the curves, it was possible to obtain parameters related to the corrosion of steels, such as corrosion potential (E_{corr}) and polarisation resistance (R_p), which are important for the evaluation of the corrosion data. Anode active regions in which the current density increases with increasing potential are observed inFigure 5 and are consistent with the behaviour of steels. For steel AISI 316, this region begins at 0.15 V vs. Ag/AgCl$_{(KCl,\ 3\ mol/L)}$, and for AISI 1010 steel, it begins at −0.45 V vs. Ag/AgCl$_{(KCl,\ 3mol/L)}$.

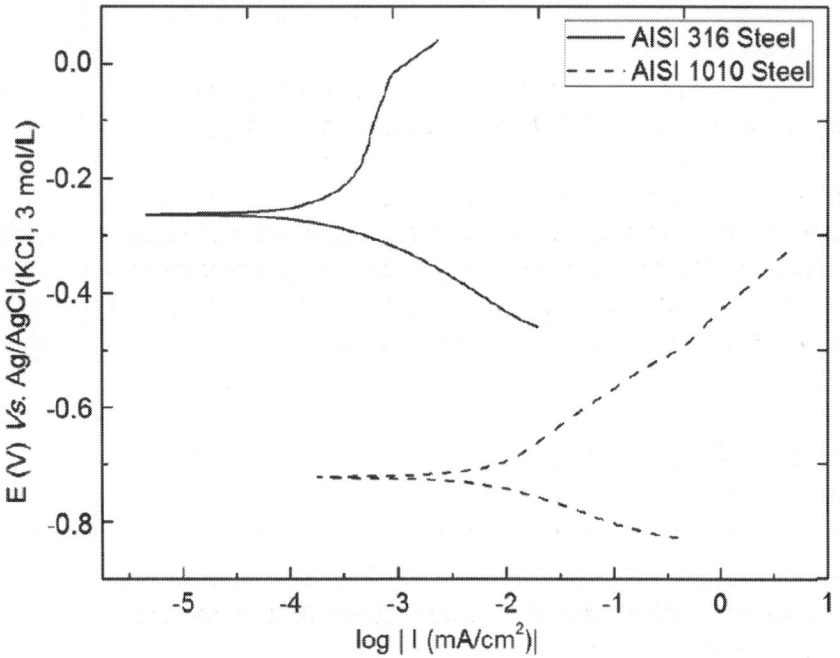

Figure 5. Tafel plots of the AISI 316 stainless steel and AISI 1010 carbon steel in the presence of 3.0 wt% solution of chloride ions. Temperature = 24°C ± 1.0°C; scan rate of 0.5 mV·s⁻¹.

Furthermore, the current density increases more sharply with the increase of the potential for AISI 1010 steel, indicating that this steel has less resistance to corrosion steel AISI 316. This finding is in agreement with the results from the literature showing that the composition of steel influences the corrosion behaviour [14] [15] . This active behaviour is caused by the dissolution of the steel in the presence of chloride ion. The steel reacts to form the ferrous chloride, which is hydrolysed to form ferrous hydroxide and hydrochloric acid. This process decreases the pH of the medium and accelerates the corrosion mechanism [41] .

The curves also show a region of potential where capacitive capacitive current exists, which is caused by the charging of the electric double layer. In this region, there is no increase in current density with increasing potential and virtually zero total current, which indicates that there is no breaking of the passive layer present on the steel surface [10] . For AISI 316, that potential range is approximately 1350 mV and is greater than that of AISI 1010 steel, which has an approximately 750 mV range. Thus, the AISI 316 steel has a more stable passive layer (data not shown).

The Tafel plots also show the corrosion resistance of AISI 316 steel compared with AISI 1010 steel, with −262 mV from E_{corr} vs. Ag/AgCl$_{(KCl, 3 mol/L)}$ and −722 mV vs. Ag/AgCl$_{(KCl, 3 mol/L)}$, respectively, in accordance with the analysis of potentiodynamic curves. This behaviour is due to the composition of each steel and the corrosion resistance imparted to the AISI 316 steel because of the presence of elements such as chromium and nickel, which are responsible for reducing the susceptibility of steel corrosion due to the possibility of film formation on the steel surface [14] [15] .

Electrochemical Study of Steels with Benzimidazole Inhibitor

Tafel curves for the AISI 316 and AISI 1010 steels in 3.0 wt% chloride ion solution + benzimidazole inhibitor at different concentrations (from 25 ppm to 1000 ppm) are shown in Figure 6. By extrapolating the Tafel plots, values of the electrochemical parameters associated with corrosion processes, such as corrosion potential (E_{corr}), corrosion current density (i_{corr}) and polarisation resistance (R_p), were obtained. The results of the corrosion inhibition efficiency of the benzimidazole were calculated from the values of the polarisation resistance (R_p) using Equation (1), as shown in the literature [42] .

$$n(\%) = \frac{Rp(\text{inhibitor}) - Rp}{Rp(\text{inhibitor})} \times 100$$

(1)

R_p (inhibitor) and R_p are the polarisation resistance values in the presence or absence of inhibitor, respectively. Table 2 lists the obtained parameters and the corrosion inhibition efficiency (n) of benzimidazole.

Table 2 shows that there was an increase in R_p for all concentrations of benzimidazole for both steels, which means that the inhibitor is inhibiting the corrosion reaction. Moreover, the corrosion potential (E_{corr}) shifted to more anodic potentials. For both steels, the best inhibition efficiency was observed at an inhibitor concentration of 100 ppm, with an efficiency of 71.4% and R_p = 8.93 kΩ·cm^2 for AISI 316 and an efficiency of 51.2% and R_p = 178.60 Ω·cm^2 for the AISI 1010 steel.

Khaled [27] used benzimidazole as a corrosion inhibitor for iron in acidic medium and showed that at a concentration of 0.05 mol·L^{-1} benzimidazole, which is greater than the concentration used in this paper, the maximum efficiency was 51%. This result shows that the benzimidazole effectively acted as an inhibitor for both the steel and the AISI 316 and AISI 1010 steels in corrosive environment studied. Tang et al. [42] used benzimidazole derivatives, such as 2-aminomethyl benzimidazole, as corrosion inhibitors for mild carbon steel in acid medium and showed a maximum corrosion inhibition efficiency of 74% at the highest concentration studied (0.002 mol·L^{-1}). Because the molecular structure of this inhibitor provides more adsorption sites compared with benzimidazole and is a molecule of greater steric body that can better cover the surface of the steel, it should be noted that benzimidazole showed a maximum inhibition efficiency of 71.4% for AISI 316 steel, which is very close to the maximum obtained by Tang et al. [42] . As observed in this study, benzimidazole showed a better inhibition for AISI 316 steel, and this behaviour was observed in almost all of the inhibitor concentrations studied.

According to the literature, benzimidazole derivatives protect the steel surface by adsorption to form an insoluble complex with Fe (II) and the segments of these derivatives that are longer than benzimidazole increase the molecule inhibition efficiency [42] . The adsorption mechanism is due to the lone pair electrons and the ϖ electrons of the aromatic ring that are present in the inhibitor molecule and occupy the empty d orbitals of Fe atoms, forming a protective film on the steel surface [43] [44] . It is also possible that inhibitor packaging products of corrosion may hinder the diffusion of aggressive species to the steel surface [42] . Such phenomena may have been responsible for the results of corrosion inhibition by benzimidazole. However, the lower inhibition efficiency obtained for AISI 1010 steel compared with AISI 316 steel suggests that the benzimidazole may have failed to adsorb to the surface of this steel, which indicates a less effective and/or weaker adsorption onto the surface. Another suggestion may be the resulting geometry of the packing of inhibitor molecules, which leaves portions of the surface exposed to the corrosive environment. Additionally, the smaller steric body of this molecule may not have protected the entire surface. For example, in studies conducted by Khaled and Amin [43] using the thiazole derivative 2-Amino-4-(p-tolyl) thiazole, which has a larger steric body than benzimidazole, the maximum inhibition efficiency was 95.2% for carbon steel.

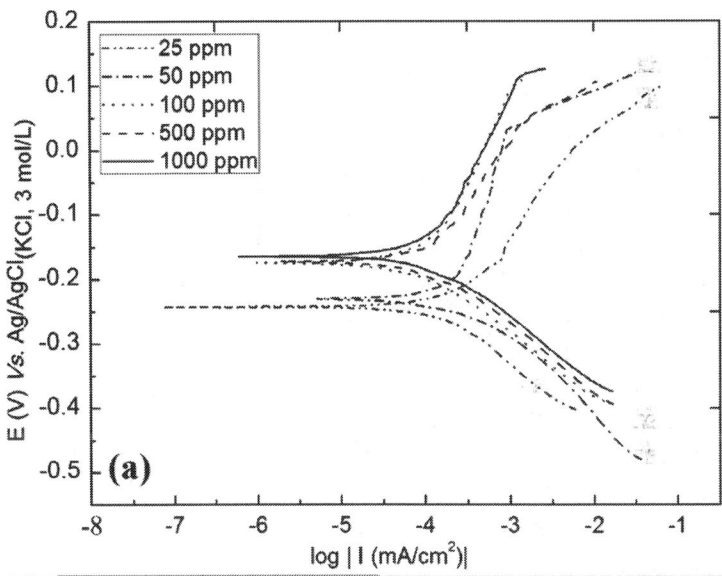

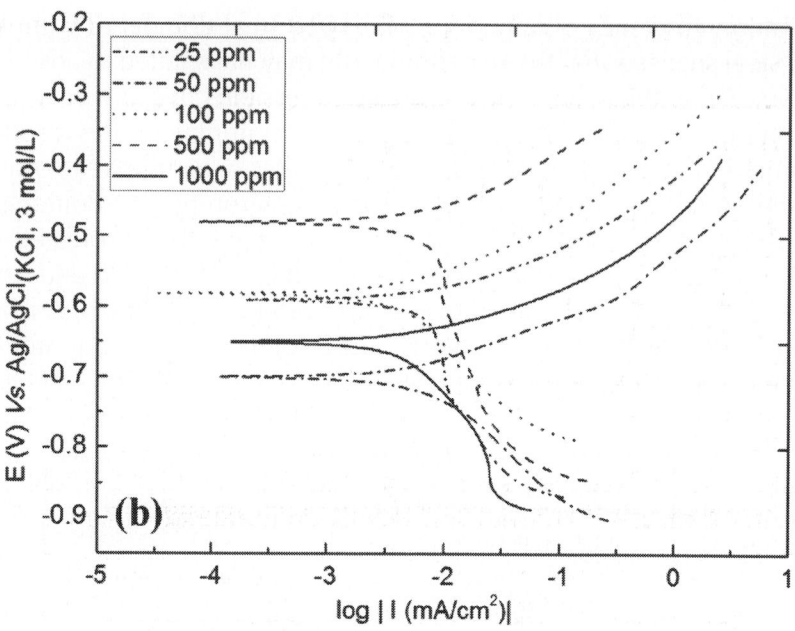

Figure 6. Tafel plots of steel in the middle of chloride ion solution at 3.0 wt% 24°C ± 1.0°C at different concentrations of benzimidazole. (a) AISI 316 stainless steel and (b) AISI 1010 carbon steel.

Table 2. Results obtained from Tafel plots for AISI 316 stainless steel and AISI 1010 carbon steel immersed in 3.0 wt% chloride ion solution at 24°C ± 1.0°C in the presence of different concentrations of benzimidazole (from 0 ppm to 1000 ppm)

Benzimida zole	AISI 316			AISI 1010		
	E_{corr} (mV)	Rp (kΩ·cm²)	n (%)	E_{corr} (mV)	Rp (Ω·cm²)	n (%)
0 ppm	-243	2.55	-	-722	87.14	-
25 ppm	-262	4.28	40.42	-591	89.29	2.41
50 ppm	-231	3.78	32.54	-700	142.90	39.02
100 ppm	-174	8.93	71.44	-582	178.60	51.21
500 ppm	-173	7.15	64.34	-480	110.50	21.14
1000 ppm	-165	5.53	53.89	-650	94.58	7.87

Electrochemical Study of Steels with Imidazole Inhibitor

Figure 7 shows the Tafel plots of AISI 316 and AISI 1010 steels obtained from the solution of 3.0 wt% chloride ions in the presence of imidazole inhibitor at different concentrations (from 25 ppm to 1000 ppm). The Tafel plots show that there is a decrease in the current density in the presence of imidazole inhibitor. Moreover, there was a change of corrosion potential (E_{corr}) to more anodic values.

The obtained parameters and the corrosion inhibition efficiency (n) of imidazole are shown in Table 3. For both steels, the inhibitor concentration that showed the best inhibition efficiency was 50 ppm, with an efficiency of ~96%, and R_p = 63.18 k$\Omega\cdot$cm^2 for the AISI 316 stainless steel and an efficiency of ~73% and R_p = 325.10 $\Omega\cdot$cm^2 for the AISI 1010 carbon steel.

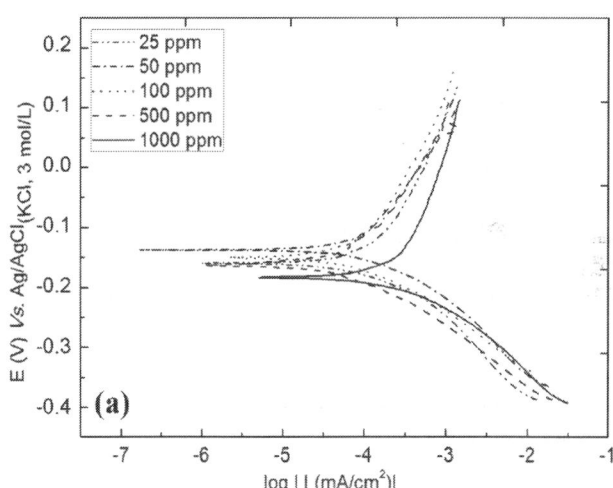

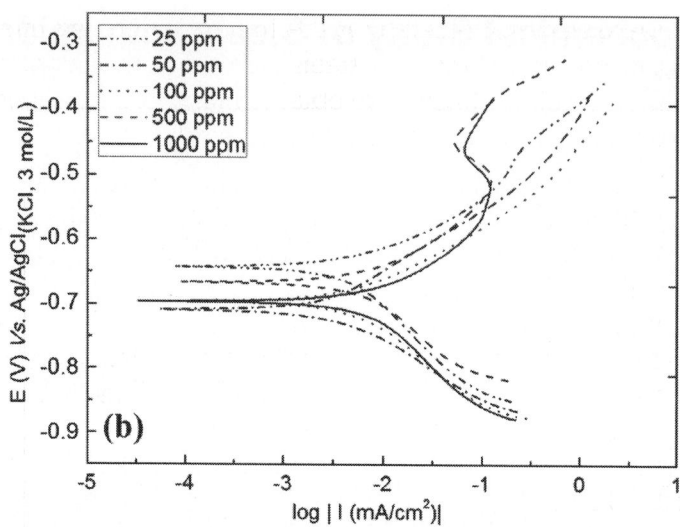

Figure 7. Tafel plots of steel in 3.0 wt% chloride ion solution at 24˚C ± 1.0˚C with different concentrations of imidazole. (a) AISI 316 stainless steel and (b) AISI 1010 carbon steel.

Table 3. Results obtained from Tafel plots for AISI 316 and AISI 1010 steels immersed in 3.0 wt% chloride ion solution at 24˚C ± 1.0˚C in the presence of different concentrations of imidazole (from 0 ppm to 1000 ppm)

Benzimid zole	AISI 316			AISI 1010		
	E_{corr} (mV)	Rp (kΩ·cm²)	n (%)	E_{corr} (mV)	Rp (Ω·cm²)	n (%)
0 ppm	-262	2.55	-	-722	87.14	-
25 ppm	-160	15.96	84.02	-643	187.20	53.45
50 ppm	-138	63.18	95.96	-709	325.10	73.20
100 ppm	-150	13.60	81.25	-696	176.70	50.68
500 ppm	-154	18.92	86.52	-666	101.00	13.72
1000 ppm	-183	3.02	15.56	-696	113.40	23.16

Zhang et al. [31] , using imidazole as a corrosion inhibitor for iron in acidic medium, achieved a maximum efficiency of 79.6% at a concentration of 0.1 mol·L⁻¹ inhibitor, which was higher than the concentration used in this study here. Another study by He et al. [45] that used imidazole to inhibit corrosion in aluminium found a maximum efficiency of 79.4% at an imidazole concentration of 0.014

mol·L^{-1}. These results indicate that the imidazole proved effective in inhibiting corrosion for the two steels studied in this work and that higher inhibition efficiencies were obtained for AISI 316 stainless steel compared with AISI 1010 carbon steel at almost all of the inhibitor concentrations studied.

The results showed that imidazole showed better corrosion inhibition efficiency than did benzimidazole. The imidazole molecule has a planar aromatic ring containing two N heteroatoms. According to Mendes et al. [36] , this molecule can be adsorbed by the imidazole ring, parallel to the surface of the metallic material, which may be one reason for the better efficiency of inhibition. Thus, it is suggested the imidazole molecules covered more of the metal surface due to a better compression between them, thus resulting in higher corrosion inhibition efficiency.

CONCLUSIONS

The analysis performed using XRD showed that the different phases are present in the steels and affect corrosion resistance. The SEM and EDX techniques make it possible to identify and determine the composition of the regions that are more susceptible to corrosion, such as inclusions in the steels. These techniques also allow for the further characterisation of the type of corrosion suffered by each steel.

Electrochemical studies have shown that at any inhibitor concentration (25 ppm to 1000 ppm), there was an increase of the polarisation resistance of both steels, indicating that there was an increase in corrosion resistance.

The best corrosion inhibition result for AISI 316 steel and AISI 1010 steel using benzimidazole was obtained at an inhibitor concentration of 100 ppm, and the inhibition efficiency values were ~71% and ~51%, respectively. When imidazole was used as the inhibitor, the best corrosion inhibition was obtained for both steels using imidazole at an inhibitor concentration of 50 ppm, with inhibition efficiency values of ~96% and ~73%, respectively.

ACKNOWLEDGEMENTS

The authors thank FAPES N° 61612626/2013, CAPES, CNPq, UFES, PETROBRAS, IFES, and LPT-UFES TRICORRMAT-UFES.

REFERENCES

1.   Groysman, A. and Brodsky, N. (2006) Corrosion and Quality. Accreditation and Quality Assurance, 10, 537-542. http://dx.doi.org/10.1007/s00769-005-0034-3

2.   Almeida, M. (2005) Minimisation of Steel Atmospheric Corrosion: Updated Structure of Intervention. Progress in Organic Coatings, 54, 81-90.http://dx.doi.org/10.1016/j.porgcoat.2005.01.007

3.   Hao, L., Zhang, S., Dong, J. and Ke, W. (2011) Atmospheric Corrosion Resistance of MnCuP Weathering Steel in Simulated Environments. Corrosion Science, 53, 4187-4192.http://dx.doi.org/10.1016/j.corsci.2011.08.028

4.   Shifler, D.A. (2005) Understanding Material Interactions in Marine Environments to Promote Extended Structural Life. Corrosion Science, 47, 2335-2352.http://dx.doi.org/10.1016/j.corsci.2004.09.027

5.   Kage, I., Matsui, K. and Kawabata, F. (2005) Minimum Maintenance Steel Plates and Their Application Technologies for Bridge-Life Cycle Cost Reduction Technologies with Environmental Safeguards for Preserving Social Infrastructure Assets. JFE Technical Report, 5, 31-37.

6.   Tsutsumi, Y., Nishikata, A. and Tsuru, T. (2007) Pitting Corrosion Mechanism of Type 304 Stainless Steel under a Droplet of Chloride Solutions. Corrosion Science, 49, 1394-1407.http://dx.doi.org/10.1016/j.corsci.2006.08.016

7.   Laycock, N.J. and Newman, R.C. (1997) Localised Dissolution Kinetics, Salt Films and Pitting Potentials. Corrosion Science, 39, 1771-1790. http://dx.doi.org/10.1016/S0010-938X(97)00049-8

8.   Newman, R.C. (2001) Understanding the Corrosion of Stainless Steel. Corrosion, 57, 1030-1041. http://dx.doi.org/10.5006/1.3281676

9.   Martini, E.M.A. and Muller, I.L. (2000) Characterization of the

Film formed on Iron in Borate Solution by Electrochemical Impedance Spectroscopy. Corrosion Science, 42, 443-454. http://dx.doi.org/10.1016/S0010-938X(99)00064-5

10. Pistorius, P.C. and Burstein G.T. (1992) Metastable Pitting Corrosion of Stainless Steel and the Transition to Stability. Philosophical Transactions of the Royal Society A, 341, 531-559. http://dx.doi.org/10.1098/rsta.1992.0114

11. Kondo, Y. (1989) Prediction of Fatigue Crack Initiation Life Based on Pit Growth. Corrosion, 45, 7-11. http://dx.doi.org/10.5006/1.3577891

12. Turnbull, A., Mccartney, L.N. and Zhou, S. (2006) A Model to Predict the Evolution of Pitting Corrosion and the Pit-to-Crack Transition Incorporating Statistically Distributed Input Parameters. Corrosion Science, 48, 2084-2105.http://dx.doi.org/10.1016/j.corsci.2005.08.010

13. Baddoo, N.R. (2008) Stainless Steel in Construction: A Review of Research, Applications, Challenges and Opportunities. Journal of Constructional Steel Research, 64, 1199-1206.http://dx.doi.org/10.1016/j.jcsr.2008.07.011

14. Malik, A.U., Siddiqi, N., Ahmad, S. and Andijani, I.N. (1995) The Effect of Dominant Alloy Additions on the Corrosion Behaviour of Some Conventional and High Alloy Stainless Steels in Seawater. Corrosion Science, 37, 1521-1535. http://dx.doi.org/10.1016/0010-938X(95)00043-J

15. Lippold, J.C. and Kotecki, D.J. (2005) Welding Metallurgy of Stainless Steel. Wiley-Intercience, New Jersey.

16. Avci, R., Davis, B.H., Wolfenden, M.L., Beech, I.B., Lucas, K. and Paul, D. (2013) Mechanism of MnS-Mediated Pit Initiation and Propagation in Carbon Steel in an Anaerobic Sulfidogenic Media. Corrosion Science, 76, 267-274.http://dx.doi.org/10.1016/j.corsci.2013.06.049

17. Rangaraju, R.R., Raja, K.S., Panday, A. and Misra, M. (2010) Low-Cost Photoelectrocatalyst Based on a Nanoporous Oxide Layer of Low-Carbon Steel. Journal of Physics D: Applied Physics, 43, Article ID: 445301. http://dx.doi.org/10.1088/0022-3727/43/44/445301

18. Ormellese, M., Lazzari, L., Goidanich, S., Fumagalli, G. and

Brenna, A. (2009) A Study of Organic Substances as Inhibitors for Chloride-Induced Corrosion in Concrete. Corrosion Science, 51, 2959-2968. http://dx.doi.org/10.1016/j.corsci.2009.08.018

19. Fekry, A.M. and Ameer, M.A. (2010) Corrosion Inhibition of Mild Steel in Acidic Media Using Newly Synthesized Heterocyclic Organic Molecules. International Journal of Hydrogen Energy, 35, 7641-7651. http://dx.doi.org/10.1016/j.ijhydene.2010.04.111

20. Lunarska, E. and Chernyayeva, O. (2006) Effect of Corrosion Inhibitors on Hydrogen Uptake by Al from NaOH Solution. International Journal of Hydrogen Energy, 31, 285-293. http://dx.doi.org/10.1016/j.ijhydene.2005.04.051

21. Zucchi, F., Trabanelli, G., Frignani, A. and Zucchini, M. (1978) The Inhibition of Stress Corrosion Cracking of Stainless Steels in Chloride Solutions. Corrosion Science, 18, 87-95. http://dx.doi.org/10.1016/S0010-938X(78)80078-X

22. Galal, A., Atta, N.F. and Al-Hassan, M.H.S. (2005) Effect of Some Thiophene Derivatives on the Electrochemical Behaviour of AISI 316 Austenitic Stainless Steel in Acidic Solutions Containing Chloride Ions I. Molecular Structure and Inhibition Efficiency Relationship. Materials Chemistry and Physics, 89, 38-48. http://dx.doi.org/10.1016/j.matchemphys.2004.08.019

23. Goudarzi, N. and Farahani, H. (2014) Investigation on 2-Mercaptobenzothiazole Behaviour as Corrosion Inhibitor for 316-Stainless Steel in Acidic Media. Anti-Corrosion Methods and Materials, 61, 20-26. http://dx.doi.org/10.1108/ACMM-11-2012-1223

24. Granero, M.F.L., Matai, P.H.L.S., Aoki, I.V. and Guedes, I.C. (2009) Dodigen 213-N as Corrosion Inhibitor for ASTM 1010 Mild Steel in 10% HCl. Journal of Applied Electrochemistry, 39, 1199-1205. http://dx.doi.org/10.1007/s10800-009-9785-6

25. Duda, Y., Govea-Rueda, R., Galicia, M., Beltrán, H.I. and Zamudio-Rivera, L.S. (2005) Corrosion Inhibitors: Design, Performance and Computer Simulations. The Journal of Physical Chemistry B, 109, 22674-22684. http://dx.doi.org/10.1021/jp0522765

26. Kuznetsov, Y.I. and Kazansky, L.P. (2008) Physicochemical Aspects of Metal Protection by Azoles as Corrosion Inhibitors. Russian Chemical Reviews, 77, 219-232. http://dx.doi.org/10.1070/

RC2008v077n03ABEH003753

27. Khaled, K.F. (2003) The Inhibition of Benzimidazole Derivatives on Corrosion of Iron in 1M HCl Solutions. Electrochimica Acta, 48, 2493-2503. http://dx.doi.org/10.1016/S0013-4686(03)00291-3

28. Wang, L. (2001) Evaluation of 2-Mercaptobenzimidazole as Corrosion Inhibitor for Mild Steel in Phosphoric Acid. Corrosion Science, 43, 2281-2289. http://dx.doi.org/10.1016/S0010-938X(01)00036-1

29. Popova, A., Christov, M. and Deligeorgiev, T. (2003) Influence of the Molecular Structure on the Inhibitor Properties of Benzimidazole Derivatives on Mild Steel Corrosion in 1 M Hydrochloric Acid. Corrosion, 59, 756-764. http://dx.doi.org/10.5006/1.3277604

30. Aljourani, J., Raeissi, K. and Golozar, M.A. (2009) Benzimidazole and Its Derivatives as Corrosion Inhibitors for Mild Steel in 1M HCl Solution. Corrosion Science, 51, 1836-1843. http://dx.doi.org/10.1016/j.corsci.2009.05.011

31. Zhang, Z., Chen, S., Li, Y., Li, S. and Wanga, L. (2009) A Study of the Inhibition of Iron Corrosion by Imidazole and Its Derivatives Self-Assembled Films. Corrosion Science, 51, 291-300. http://dx.doi.org/10.1016/j.corsci.2008.10.040

32. Curkovic, H.O., Stupnišek-lisac, E. and Takenouti, H. (2010) The Influence of pH Value on the Efficiency of Imidazole Based Corrosion Inhibitors of Copper. Corrosion Science, 52, 398-405. http://dx.doi.org/10.1016/j.corsci.2009.09.026

33. Zhang, D.Q., Gao, L.X. and Zhou, G.D. (2004) Inhibition of Copper Corrosion in Aerated Hydrochloric Acid Solution by Heterocyclic Compounds Containing a Mercapto Group. Corrosion Science, 46, 3031-3040. http://dx.doi.org/10.1016/j.corsci.2004.04.012

34. Popova, A., Sokolova, E., Raicheva, S. and Christov, M. (2003) AC and DC Study of the Temperature Effect on Mild Steel Corrosion in Acid Media in the Presence of Benzimidazole Derivatives. Corrosion Science, 45, 33-58. http://dx.doi.org/10.1016/S0010-938X(02)00072-0

35. Kokalj, A. (2013) Formation and Structure of Inhibitive Molecular Film of Imidazole on Iron Surface. Corrosion Science, 68, 195-203. http://dx.doi.org/10.1016/j.corsci.2012.11.015

36. Mendes, J.O., Silva, E.C. and Rocha, A.B. (2012) On the Nature of Inhibition Performance of Imidazole on Iron Surface. Corrosion Science, 57, 254-259.http://dx.doi.org/10.1016/j.corsci.2011.12.011

37. Mousavi, M., Mohammadalizadeh, M. and Khosravan, A. (2011) Theoretical Investigation of Corrosion Inhibition Effect of Imidazole and Its Derivatives on Mild Steel Using Cluster Model. Corrosion Science, 53, 3086-3091. http://dx.doi.org/10.1016/j.corsci.2011.05.034

38. ASM International Handbook Committee (1990) Volume 1: Properties and Selection: Irons, Steels and High-Perfor- mance Alloys. ASM International, Ohio, 150.

39. Li, Y., Hu, R.G., Wang, J.R., Huang, Y.X. and Lin, C.-J. (2009) Corrosion Initiation of Stainless Steel in HCl Solution Studied Using Electrochemical Noise and in Situ Atomic Force Microscope. Electrochimica Acta, 54, 7134-7140.http://dx.doi.org/10.1016/j.electacta.2009.07.042

40. Gammer, K., Rosner, M., Poeckl, G. and Hutter, H. (2003) AES and SIMS Analysis of Non-Metallic Inclusions in a Low-Carbon Chromium-Steel. Analytical and Bioanalytical Chemistry, 376, 255-259. http://dx.doi.org/10.1007/s00216-003-1851-z

41. Yang, Y.Z., Jiang, Y.M. and Li, J. (2013) In Situ Investigation of Crevice Corrosion on UNS S32101 Duplex Stainless Steel in Sodium Chloride Solution. Corrosion Science, 76, 163-169. http://dx.doi.org/10.1016/j.corsci.2013.06.039

42. Tang, Y., Zhang, F., Hu, S., Cao, Z., Wu, Z. and Jing, W. (2013) Novel Benzimidazole Derivatives as Corrosion Inhibitors of Mild Steel in the Acidic Media. Part I: Gravimetric, Electrochemical, SEM and XPS studies. Corrosion Science, 74, 271-282.http://dx.doi.org/10.1016/j.corsci.2013.04.053

43. Khaled, K.F. and Amin, M.A. (2009) Corrosion Monitoring of Mild Steel in Sulphuric Acid Solutions in Presence of Some Triazole Derivatives—Molecular Dynamics, Chemical and Electrochemical Studies. Corrosion Science, 51, 1964-1975. http://dx.doi.org/10.1016/j.corsci.2009.05.023

44. Negma, N.A., Kandile, N.G., Badr, E.A. and Mohammed, M.A. (2012) Gravimetric and Electrochemical Evaluation of Environmentally Friendly Nonionic Corrosion Inhibitors for

Carbon Steel in 1 M HCl. Corrosion Science, 65, 94-103.http://dx.doi.org/10.1016/j.corsci.2012.08.002

45. He, X., Jiang, Y., Li, C., Wang, W., Hou, B. and Wu, L. (2014) Inhibition Properties and Adsorption Behaviour of Imidazole and 2-Phenyl-2-Imidazoline on AA5052 in 1.0 M HCl Solution. Corrosion Science, 83, 124-136. http://dx.doi.org/10.1016/j.corsci.2014.02.004

Chapter 7

Corrosion and Surface Treatment of Magnesium Alloys

Henry Hu[1], Xueyuan Nie[1], and Yueyu Ma[1]

[1]Department of Mechanical, Automotive and Materials Engineering, University of Windsor, Windsor, Canada

INTRODUCTION

The need for fuel efficiency and increased performance in transportation systems continually places new demands on the materials used. The design criteria which automobile and aerospace industries are primarily concerned with are density, strength, stiffness, and corrosion resistance. Low-density materials may reduce fuel costs, increase range, and allow larger payloads. High strength and stiffness are necessary for adequate performance and safety characteristics, while corrosion resistance helps to ensure that design lifetime is achieved.

Magnesium is the 8th most abundant element on the earth making up approximately 1.93% by mass of the earth's crust and 0.13% by mass of the oceans [1]. Other advantages of magnesium alloys have played an important role in a broad variety of structural applications in the automobile, aerospace, electronics, and consumer products industries. Magnesium has specific high strength to weight ratio, and it is 35% lighter than aluminium and 75% lighter than iron. Typical magnesium alloys weigh ~25% less than their aluminium counterparts at equal stiffness. Magnesium also has high thermal conductivity, good electromagnetic shielding characteristics, good ductility, excellent castability and better damping characteristics than aluminum, and Mg is easily recycled. The main use of magnesium by far is as an alloying addition to aluminum alloys. Other major uses of magnesium include desulphurization of steel and the production of ductile iron. As a structural material, it can be used in aerospace components, automobile and computer parts, mobile phones and sporting goods. Magnesium for structural applications is processed into castings (die, sand, permanent mold and investment), extrusions, forgings, impact extrusions and flat rolled products. Die castings account for 70% of the castings shipped. Magnesium can be joined by riveting, or any of the commonly used welding methods [2].

With the dramatically increased emphasis on weight reduction, magnesium is receiving a lot of attention as a material for use in the next generation automobiles. This is due to limited fossil fuel supplies and arising environmental problems associated with fuel emission products. Magnesium alloys are a promising alternative to the aluminium alloys currently dominating the transportation industry. However, the limited use of magnesium in engineering applications results mainly from the shortcomings including high reactivity in the molten state, inferior fatigue and creep properties compared to aluminium, poor corrosion and wear resistance [22]. One of the main challenges in the use of magnesium, particularly for outdoor application, is to overcome its poor corrosion resistance. Magnesium and its alloys are extremely susceptible to galvanic corrosion, which can cause severe attack in the metal resulting in decreased mechanical stability and an unattractive appearance. Corrosion can be minimized by the use of high purity alloys in which the heavy metal impurities such as iron, nickel and copper are kept below a threshold value. The elimination of bad design, surface contamination, galvanic couples and inadequate or incorrectly applied surface protection schemes can also significantly decrease the corrosion rate of magnesium alloys in service [3].

In this chapter, the corrosion characteristic of Mg and Mg alloys are described. Fundamental aspects of magnesium corrosion such as general corrosion, galvanic corrosion, pitting, stress corrosion and corrosion fatigue are reviewed. The factors that control the corrosion behaviour of Mg and Mg alloys are discussed in some detail. Finally, the more recently developed corrosion science and engineering underpinning various surface treatment methods such as electrochemical plating, conversion coating, anodizing, gas-phase coating, organic coating, electrolytic plasma oxidation for magnesium alloys are described.

CORROSION CHARACTERISTICS OF PURE MAGNESIUM

Magnesium, like most metals and alloys, relies on a natural surface film to control its corrosion. However, the nature of this film is not thoroughly understood. Good passive films are those that restrict the outward flow of cations, resist the inward flow of damaging anions or oxidants, and rapidly repair themselves in the event of localized breakdown. The structure and composition of the surface films, which depends strongly on environmental and metallurgical factors, such as electrolyte species and impurities in the metal, determine the protective ability of a passive film.

Environmental Effects

No material shows high corrosion resistance in all kinds of environments. The high corrosion resistance of materials always refers to some specific environments. Magnesium has its own preferred service environments. However, there are fewer media that are suitable for the magnesium and magnesium alloys compared with other materials, such as steels and aluminum alloys. For example, magnesium and magnesium alloys are usually stable in basic solutions, but in neutral and acidic media they dissolve at high rates [3]. This is quite different from aluminum alloys that are normally stable in neutral media but are unstable in both basic and acidic solutions.

General Corrosion in Aqueous Solutions

With few exceptions, there is no appreciable corrosion of pure magnesium near room temperature unless water is present [4]. Magnesium dissolution in water or aqueous environments generally proceeds by an electrochemical reaction with water to produce magnesium hydroxide and hydrogen gas. Such a mechanism is relatively insensitive to the oxygen concentration, although the presence of oxygen is an important factor in atmospheric corrosion [5]. Reaction 1 describes the probable overall reaction:

$$Mg + 2H_2O = Mg(OH)_2 + H_2$$

(1)

This net reaction can be expressed as the sum of the following partial reactions:

Anodic reaction : $Mg \rightarrow Mg^{2+} + 2e$

(2)

Cathodic reaction : $2H_2O + 2e \rightarrow H_2 + 2OH^-$

(3)

Products formation : $Mg^{2+} + 2OH^- \rightarrow Mg(OH)_2$

(4)

The reduction process of hydrogen ions and the hydrogen overvoltage of the cathode play an important role in the corrosion of Mg. Low overvoltage cathodes facilitate hydrogen evolution, causing a substantial corrosion rate [6].

Fig. 1 shows the corrosion domains of Mg in the Mg-H$_2$O system. The region of water stability lies between the line a and line b. At a potential below line a, hydrogen is evolved; above line b, oxygen is evolved. The numbers identify the reactions that separate the different phases shown in reactions 5, 6, and 7. The horizontal and vertical parallel lines for reactions 5 and 6 give the concentration of Mg^{2+} in mol l^{-1} as a power of 10. As shown in Fig. 1, the ringed numbered lines separate the regions of corrosion (dissolved cations, e.g. Mg^{2+}), immunity (unreacted metal, Mg), and passivation (corrosion products, $Mg(OH)_2$) [5]. From Fig. 1, it can be seen that stable films would be expected to form depending on the values of the potential and pH. In neutral and alkaline environments, the magnesium hydroxide product can form a surface film that offers considerable corrosion protection to the pure magnesium or its common alloys, although this is not as effective as the oxide layer formed on aluminum. As corrosion proceeds, the metal surface experiences a local pH increase because of the formation of $Mg(OH)_2$, whose equilibrium pH is about 11. The protection supplied by this film is therefore highly dependent on the condition of exposure. High purity magnesium is reported to have a corrosion rate of 10^{-2}-10^{-3} mils per year (mpy) when exposed to 2 normal KOH solutions at 25 °C [3].

Linea: $2H^+ + 2e \rightarrow H_2$

Lineb: $4OH^- - 4e \rightarrow O_2 + H_2O$

1. $Mg + 2H_2O = Mg(OH)_2 + H_2$

$$(5)$$

2. $Mg^{2+} + H_2O = MgO + 2H^+$

$$(6)$$

3. $Mg = Mg^{2+} + 2e$

$$(7)$$

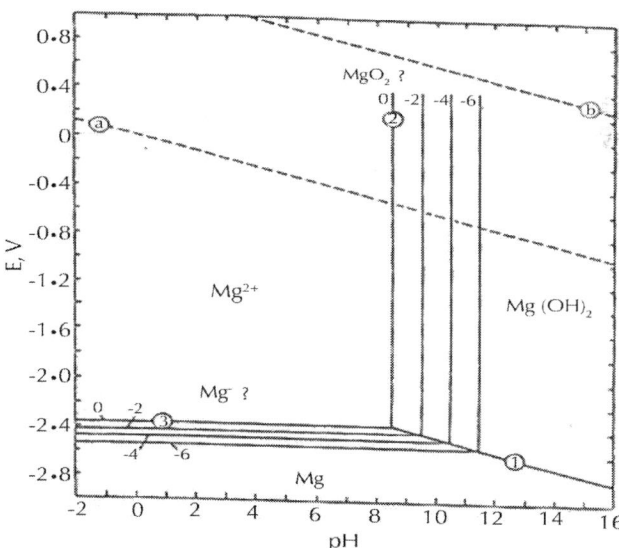

Figure 1: Electrochemical potential-pH equilibrium diagram for the magnesium-water system at 25°C [5].

Magnesium's corrosion performance in pure water is strongly dependent on temperature. At elevated temperatures, the resistance to corrosion in water decreases with increasing temperature, corrosion becoming particularly severe above 100°C [3].

Magnesium is subject to dissolution by most acids. Even in dilute solutions of strong and moderately weak acids, magnesium dissolves rapidly. There are a few exceptions, such as chromic acid and hydrofluoric acid [6]. Very slow dissolution of magnesium in chromic acid is due to its becoming passive in this acid. An insoluble surface film of MgF_2 is formed which protects against further attack, is the reason why magnesium is resistant to hydrofluoric acid [66].

The strong alkalinity of the natural hydroxide film on magnesium means there is little tendency for the compound to give up a proton to strong alkalis; consequently, the film provides excellent protection even in strong hot alkali solutions that would readily attack aluminum or zinc alloys [6, 7]. Magnesium's resistance to alkali attack combined with the metal's lightweight has made it the preferred material for cement finishing tools for many years [7].

Corrosion in the Solutions Containing Specific Ions

Salt solutions vary in their corrosivity to magnesium [7-9]: alkali metal or alkaline-earth metal (chromates, fluorides, phosphates, silicates, vanadates, or nitrates) cause little or no corrosion. Chromates, fluorides, phosphates, and silicates in particular are frequently used in the chemical treatment and anodize for magnesium surfaces due to their ability to form somewhat protective films. Chlorides, bromides, iodides and sulfates normally accelerate the corrosion of magnesium in aqueous solutions. Practically all heavy metal salts are likely to cause corrosion since magnesium normally displaces heavy metals from solution due to its high chemical activity, except iron phosphate solution.

Song et. al [10] investigated the electrochemical corrosion of pure magnesium in 1N NaCl and Na_2SO_4 solutions. It was found that a partially protective surface film plays an important role in the electrochemical dissolution processes for magnesium in NaCl, and Na_2SO_4 solutions. The presence of Cl·made the surface films more active or increased the broken area of the naturally-formed protective film, and also accelerated the electrochemical reaction rate from magnesium to univalent ions according to the reactions 8 and 9, thus increasing the corrosion rates. SO_4^- has less effect than Cl^-.

$$Mg \leftrightarrow Mg^+ + e \tag{8}$$

$$Mg^+ + H_2O \rightarrow Mg^{2+} + OH^- + \left(\frac{1}{2}\right) H_2 \tag{9}$$

In a review given by Makar and Kruger [17], it is revealed that films on magnesium immersed in 3% sodium chloride consist of $Mg(OH)_2$, $MgCl_2 \bullet 6H_2O$ and $Mg_3(OH)_5Cl \bullet 4H_2O$, which were identified by infrared spectroscopy and X-ray diffractions. In addition, the stabilized film on magnesium also includes MgH_2.

Oxidizing anions, especially chromates, dichromates, and phosphates, which form protective films, can strongly increase the corrosion resistance of magnesium in water or aqueous salt solutions [6].

Corrosion Caused by Organic Compounds

In a review [7], it was revealed that organic compounds, with a few exceptions, have little effect on magnesium and its alloys. It has been indicated in references 7 and 9 that magnesium is usable in contact with aromatic and aliphatic hydrocarbons, ketones, esters, ethers, glycols, phenols, amines, aldehydes, oils, and higher alcohols. Ethanol causes slight attack, but anhydrous methanol causes severe attack unless significant water content is introduced. Most dry chlorinated hydrocarbons cause little attack on magnesium up to their boiling points. In the presence of water, particularly at high temperatures, chlorinated hydrocarbons may hydrolyze to form hydrochloric acid, causing corrosive attack of the magnesium. Dry fluorinated hydrocarbons, for example, refrigerants, do not attack magnesium at room temperature. When water is present, however, hydrolysis may cause corrosive attack. In acidic food stuffs, such as fruit juices and carbonated beverages, attack of magnesium is slow but measurable. Milk causes attack, particularly when souring [7, 9].

A magnesium engine block has been targeted for reducing the weight of an automobile and corrosion is a major issue in the cooling system of an engine block. It was reported by Song et al [11] that some inhibitors in the traditional coolants, whose main compositions is 30-70 vol% ethylene glycol and molybdate, phosphate, borate, nitrite, tolyltriazole, benzoate and silicate inhibitors, fail to provide adequate corrosion protection to magnesium and magnesium alloys. Hence, some companies are developing coolants with new inhibitors for magnesium and magnesium alloys. Song et al. [11] studied the corrosion behaviour of pure magnesium in ethylene glycol containing various ions. It was found that the corrosion rate of magnesium in aqueous ethylene glycol depends on the concentration of the solution. A dilute ethylene glycol solution is more corrosive than a concentrated solution at room temperature. An ethylene glycol solution contaminated by individual contaminants NaCl, $NaHCO_3$ and Na_2SO_4 is more corrosive to pure magnesium. NaCl is the most detrimental contaminate, while in a NaCl contaminated ethylene glycol solution, a small amount of $NaHCO_3$ or Na_2SO_4 has some inhibition effect [11]. Fluorides in ethylene glycol can effectively reduce the corrosion of magnesium due to the formation of a protective fluoride-containing film on the magnesium surface. It has been observed [11] that a small amount of contaminants (Na_2SO_4, $NaHCO_3$) addition decreases the corrosion of magnesium in the chloride containing ethylene glycol solution.

Corrosion in the Air

Humidity plays a major role in the corrosion of magnesium [12]. Corrosion of magnesium increases with relative humidity. At 10% humidity, pure magnesium does not show evidence of surface corrosion after 18 months. However, at 30% humidity, a small amount of visible surface oxide haze and slight corrosion is evident, while at 80% humidity, an amorphous phase is clearly present over about 30% of the surface and the surface exhibits considerable corrosion. Crystalline magnesium hydroxide is formed only when relative humidity is at or above 93%. A theoretical explanation about less ordered films providing better protection was presented in reference 5 that a film without grain boundaries resists the movement of ions better than a crystalline film.

Furthermore, the presence of 300 ppm CO_2 and normally 1 ppm of SO_2 in the atmosphere also plays an important role in a formation of the surface films. In the atmosphere, an inhibitive effect of CO_2 in humid air has been reported [4]. Initially, the ambient levels of carbon dioxide enhance the corrosion attack, however, the rate of corrosion in the presence of CO_2 decreases with increased exposure time. It is suggested that the initial enhance of corrosion stems from the protolysis of carbonic acid, causing a pH decrease in the surface electrolyte as can be seen in reactions 10 and 11. The reduced pH in the surface electrolyte acts to increase the rate of dissolution of the air-formed film.

$$CO_2(aq) + H_2O \leftrightarrow HCO^-_3 + H^+ \tag{10}$$

$$HCO^-_3 \leftrightarrow CO_3^{2-} + H^+ \tag{11}$$

The hydroxide ions, generated in the cathodic reaction or dissolved from the film, can form carbonate with carbonic acid. In the presence of CO_2, a magnesium hydroxyl carbonate is formed (reaction 12).

$$2Mg^{2+} + CO_3^{2-} + 2OH^- + 3H_2O = Mg_2(OH)_2 CO_3 \cdot 3H_2O \tag{12}$$

Magnesium hydroxyl carbonate may also form by reaction of solid magnesium hydroxide with CO_2 and water. The presence of the carbonate film, which is thicker than the magnesium hydroxide film, interferes with both the anodic and the cathodic reaction and thus reduces the corrosion rate. Further, the protolysis of CO_2 counteracts the development of pH gradients on the surface, impeding the development of macroscopic corrosion cells, resulting in inhibition of pitting corrosion. Hence, high purity magnesium and magnesium alloys have the potential to be extremely corrosion resistant, and perform better in the atmosphere than iron.

In urban/industrial locations $MgSO_4\text{-}6H_2O$ and $MgSO_3\text{-}6H_2O$ can predominate in the surface films. $MgSO_4\text{-}6H_2O$ and $MgSO_3\text{-}6H_2O$ are highly soluble and are easily washed away, re-exposing the surface. Hence, pure magnesium has a poor corrosion resistance in industrial atmospheres [3, 9].

Metallurgical Effects

Magnesium becomes susceptible to accelerated corrosion if there are significant impurity levels present or it is in contact with other metals. Due to the lack of a nature surface film on the impurities, the more positive potential allows impurities to be efficient cathodes for hydrogen discharge, thereby providing significant microgalvanic acceleration of the corrosion rate [13]. Therefore even small amount of impurities in pure magnesium with metals having low hydrogen overvoltages, such as Fe, Ni, Co, or Cu, drastically reduces its corrosion resistance. Metals with higher hydrogen overvoltages, such as lead, zinc, and cadmium, and also strongly electronegative metals, such as manganese and aluminum, are less dangerous in this respect [6]. Fig. 2 shows effect of impurity and alloying elements on the corrosion of magnesium in a 3% NaCl solution at room temperature. Fe, Cu, Ni can increase the corrosion rate, while Cd, Pb, Sn, and Al can drastically reduce the corrosion resistance of pure magnesium. The effect of various elements on the corrosion of magnesium alloys will be discussed in detail in section 3.2.

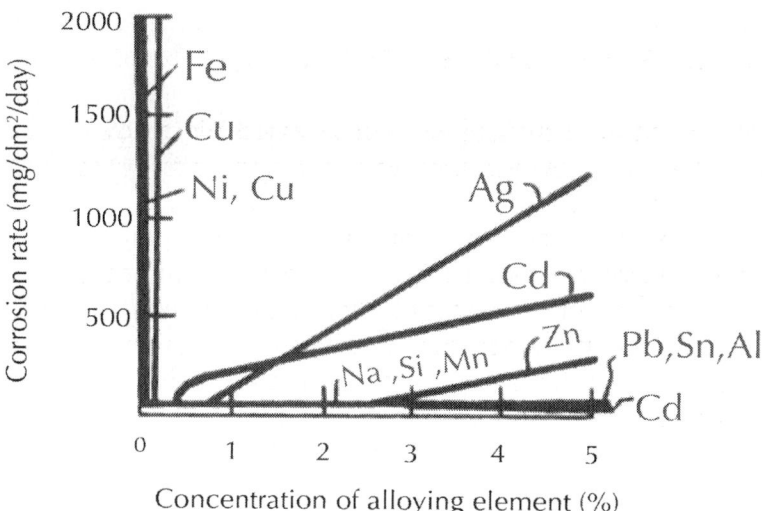

Figure 2: Effect of impurity and alloying elements on corrosion of magnesium (all alloys are formed by magnesium and the given element) in 3% NaCl solution at room temperature [6].

CORROSION CHARACTERISTICS OF MG ALLOYS

Magnesium alloys in general can be divided into two main groups: (1) those containing aluminium as the primary alloying element; and (2) those free of aluminium and containing small additions of zirconium for the purpose of grain refinement. The most widely used magnesium alloys are those with aluminum (to 10%), zinc (to 3%), and manganese (to 2.5%). It is desirable that other metals, particularly Fe, Cu, Ni and Si be present in very small amounts not exceeding a total of 0.4% to 0.6% [13]. Mg alloys corrosion is governed by the characteristics of its surface film. The properties of film on Mg alloys depend on Mg alloys'metallurgy and environmental factors. Magnesium metallurgy includes alloying and impurity elements, phase components and microstructure. Metallurgical manipulation can provide an effective means to improve the corrosion resistance of magnesium alloys.

Influences of Environment

Corrosion by Atmosphere and Solutions

In general, atmospheric attack in damp conditions is largely superficial. The corrosion reactions of magnesium alloys are similar to those for pure magnesium, as shown in reactions 1–4. Generally, the corrosion resistance of magnesium alloy is better than that of pure magnesium, because other corrosion resistant phases exist. An analysis of the films formed when magnesium alloys containing Al, Mn, or Zn are exposed to the atmosphere, shows an enrichment of the secondary constituents. It was suggested that the air-formed oxide on Mg-Al alloys has a layered structure composed of MgO/Mg-Al-oxide/substrate, with the Mg-rich oxide becoming thinner with increasing Al content. It is likely that this benefit of Al is related to the strong tendency for Al to form a stable passive film [13].

Lindstom et al [14] studied the influence of NaCl and CO_2 on the atmospheric corrosion of magnesium alloy AZ91. The combination of high humidity and NaCl is very corrosive towards AZ91. However,

CO_2 inhibits atmospheric corrosion both in the presence of and in the absence of NaCl. In the absence of CO_2, the NaCl-induced corrosion is localized and the main corrosion product is $Mg(OH)_2$. Because of the cathodic reaction, high pH areas develop in the electrolyte adjacent to the AZ91surface, resulting in the dissolution of alumnia in the passive film. Due to the rapid hydrogen evolution, the metal disintegrates and pieces of un-reacted metal are embedded in the corrosion product. In the presence of CO_2, AZ91 suffers general corrosion and carbonate-containing corrosion products were formed. In the presence of NaCl, $Mg_5(CO_3)_4(OH)_2$ was detected by XRD. The inhibitive effect of CO_2 was suggested to be due to a combination of pH decrease in the surface electrolyte, stabilizing alumina in the passive film, and the formation of sparingly soluble carbonate-containing corrosion products that slow down the electrochemical reactions.

Corrosion in Coolants

A magnesium alloy engine block has the potential to significantly reduce the weight of an automobile, and studies on the corrosion of magnesium alloys in automotive coolants has received increasing attention. Song and StJohn [15] studied the corrosion of a new magnesium alloys AM-SC1 and magnesium alloy AZ91 in commercial engine coolants. AM-SC1 is a new alloy with rare earth elements, zinc and zirconium as the main alloying elements, which was developed as an engine block material. It was found that AZ91 is more corrosion resistant than AM-SC1 in existing coolants, because the existing commercial coolants are non-corrosive to aluminium alloys and they also have a certain degree of inhibitive effect on an aluminium-containing alloy. Potassium fluoride, KF, was an effective inhibitor for the magnesium alloys and can reduce the general and galvanic corrosion rates, while it had no detrimental effect on the other engine block materials (such as cast iron, aluminium alloy) in terms of their corrosion performance. Furthermore, Toyota and Ford long life coolant (Toyota long life coolant: 934 ml/L ethylene and 10ml/kg denatonium benzoate, 1:1 diluted with demineralised water; Ford long life coolant: 950 ml/l ethylene glycol and 10 ppm denationium benzoate as a bittering agent, 1:1 diluted with demineralised water) [15] can be the most promising coolants for magnesium alloys engine blocks, because the corrosion rates in them were acceptable. Slavcheva et.al [16] demonstrated that Lactobiono-tallowamide (a reaction product of lactobionic acid and tallowamine) had relatively high inhibition efficiency towards AZ91 alloy corrosion in ethylene glycol solution containing chloride ions at low as well as at high temperature. The inhibiting effect of lactobiono-tallowamide is a result of adsorption on the metal surface and formation of an adherent protective film. Hence, lactobiono-tallowamide can be considered as a promising inhibitor of magnesium alloy corrosion.

Metallurgical Factors

Metallurgical factors include alloying and impurity elements, phase composition and microstructure.

Impurity Elements

Studies [12, 17-20] have confirmed that the most critical factor in the corrosion behaviour of Mg and Mg alloys is the metal purity. Iron, nickel, and copper are extremely deleterious because they have low solid-solubility limits and provide active cathodic sites which lead to galvanic corrosion and increase corrosion rates. At the same concentration, the detrimental effect of these elements decreases as follows: Ni>Fe>Cu. When the impurity concentration exceeds the tolerance limit, the corrosion rate is greatly accelerated, whereas the corrosion rate is low when the impurity concentration is lower than the tolerance limit. The tolerance limits in magnesium alloys are influenced by the presence of other elements. For example, the iron tolerance limit for magnesium-aluminum alloys depends on the Mn or Zn concentration [5]. Furthermore, impurity limits are different depending on the method of manufacture. For example, die cast AZ91 has higher nickel tolerance than gravity cast AZ91. And the slower solidification rates significantly affect the nickel tolerance, but not Fe and Cu [17]. Different alloys have different tolerance limits as summarized in Table 1. In Table 1, the impurity tolerance limit of 0.032 Mn actually means that if there is manganese in a magnesium alloy, then the alloy would be able to tolerate an amount of the iron impurity equal to 0.032 of the manganese concentration (by weight).

Table 1: Tolerance limits for magnesium and magnesium alloys [113]

Specimen	Condition	Tolerance Limits*		
		Fe	Ni	Cu
Pure Mg		170 ppm	5 ppm	1000 ppm
Pure Mg		170 ppm	5 ppm	1300 ppm
AZ91		20 ppm	12 ppm	900 ppm
AZ91		0.032 Mn	50 ppm	400 ppm
AZ91	High pressure (F)	0.032 Mn	50 ppm	400 ppm
AZ91	Low pressure (F)	0.032 Mn	10 ppm	400 ppm
AZ91	Low pressure (T4)	0.035 Mn	10 ppm	100 ppm
AZ91	Low pressure (T6)	0.046 Mn	10 ppm	400 ppm
AZ91B			<100ppm	<2500 ppm
AS41	Die casting	0.032 Mn	50 ppm	400 ppm
AZ91	Die casting	50 ppm	50 ppm	700 ppm
AZ91	Die casting	0.032 Mn	50 ppm	700 ppm
AZ91	Gravity casting	0.032 Mn	10 ppm	400 ppm
AM60	Die casting	0.021 Mn	30 ppm	10 ppm
AE42		0.01 Mn	40 ppm	200 ppm

[i] - * Variation of tolerance limits results from different manufacturing processes used by various alloy producers

Iron, nickel, copper, and cobalt are the four main elements so far found to have significant detrimental influence on the corrosion resistance of magnesium alloys [13]. Besides these detrimental elements, there is one special element, manganese, which is usually closely related to the detrimental effects of other elements and their tolerance limits.

Iron

The deleterious effect of iron in pure magnesium is shown in Fig. 3, and it is suspected to be due to the galvanic coupling between the magnesium matrix and the iron particles scattered in the matrix because Fe has a very low solid solubility in magnesium (about 9.9 ppm) [13], while Hawke and Olsen [18] thought that in the absence of

Mn, virtually all the Fe precipitates in magnesium alloys as FeAl$_3$ which has a high cathodic activity for corrosion, as shown in Fig. 4.

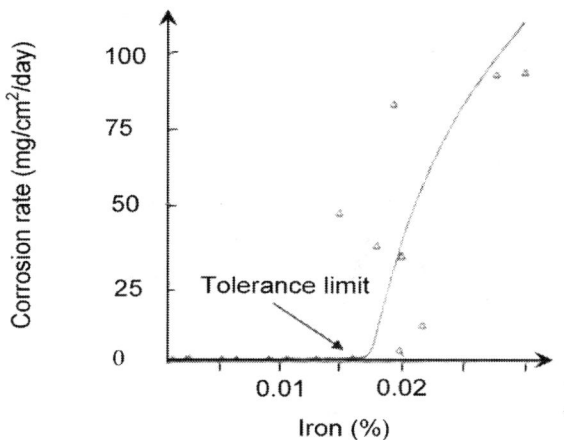

Figure 3: Effect of iron on the corrosion of pure magnesium immersed in 3% NaCl [13].

In Fig. 4, the effects of these phases are defined in terms of their electrode potential relative to the alloy matrix, and the overpotential values for the evolution of hydrogen gas. Within an appropriate medium, FeAl$_3$ acts as an effective cathode, catalyzing the reduction reaction, hydrogen evolution, which is responsible for the corrosion process. Due to the low solubility of FeAl$_3$ in Mg, increasing additions of Al result in increasingly smaller tolerance levels for Fe [17]. For example, when even as little as hundreds of ppm of aluminum is added to the magnesium, the tolerance limit for iron decreases from 170 wt.-ppm to a few wt.-ppm. With 7% Al, the tolerance is about 5 wt.-ppm Fe, while with 10% Al, the limit is too low to be determined [13].

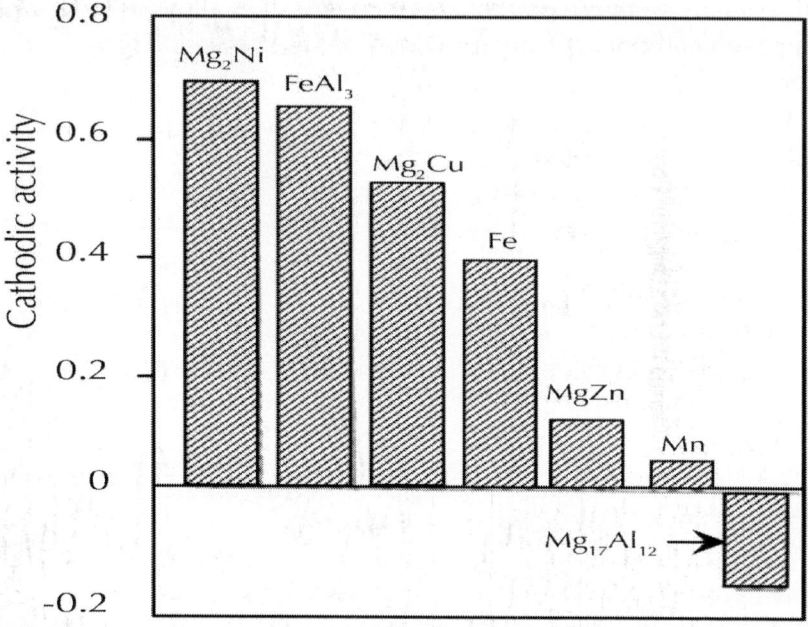

Figure 4: Cathodic activity of precipitated phases in Mg alloys in salt water relative to their alloy matrix [18].

Nickel

Nickel is more harmful than iron both in pure magnesium and in magnesium alloys, because nickel precipitates in magnesium alloys as Mg_2Ni which is more cathodically active than $FeAl_3$ and Fe, as shown in Fig. 4. Fig. 5 shows effect of nickel content on corrosion rates of pure Mg and Mg-Mn and Mg-Zn alloys using standard Dow immersion-emersion test in NaCl solution [5]. Furthermore, solidification rate changes the tolerance limit for Ni (higher tolerance limit for faster cooling, Fig. 6), but not for Fe, or Cu [17]. From Fig. 6, it was found that the nickel tolerance was significantly lower in the gravity cast (~ 10 ppm) than for the die cast (~55 ppm) [17].

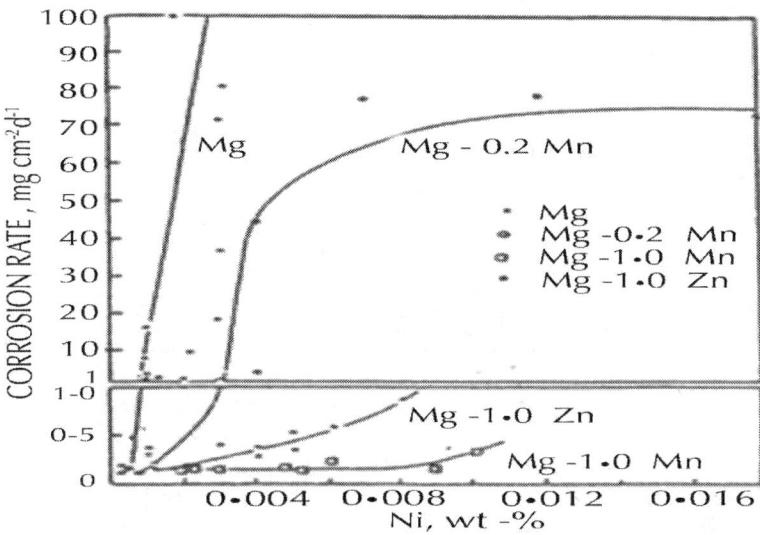

Figure 5: Effect of nickel content on corrosion rates of pure Mg and Mg-Mn and Mg-Zn alloys using standard Dow immersion-emersion test in NaCl solution [5].

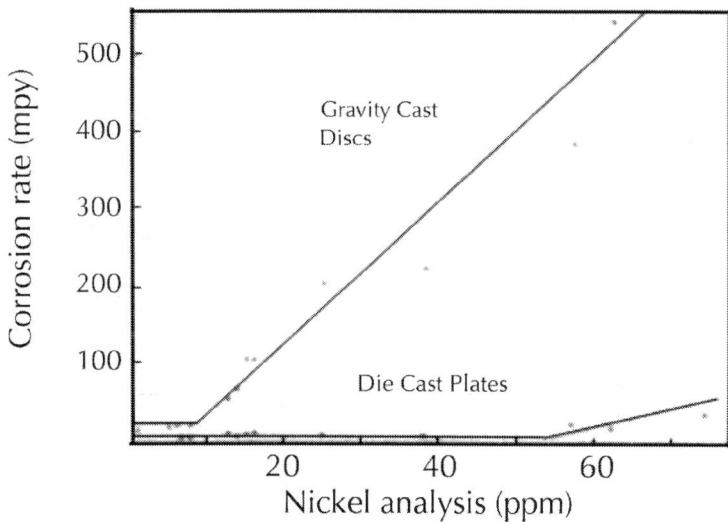

Figure 6: Salt spray corrosion performance vs nickel content for die cast and gravity cast samples [17].

Copper

A small amount of copper has a beneficial effect on the creep strength of magnesium die castings, but strongly accelerates salt water corrosion [113]. Cu is less harmful than iron in magnesium, because copper precipitates in magnesium as Mg_2Cu which has a lower potential than $FeAl_3$ [18]. Mg_2Cu acts as an effective cathode, catalyzing the hydrogen evolution. The addition of copper to Mg-Al-Zn alloys has also been shown to have a detrimental effect on the corrosion resistance. This may be attributed to the incorporation of the copper in the eutectic phase as Mg(Cu, Zn). The tolerance limit of copper has been set at 300 ppm, but it is known that higher levels can be tolerated if the zinc content is above the specification minimum of 0.4% [113].

Manganese

Manganese is added to many commercial alloys, particularly the Mg-Al-Zn alloys (AZ series) to improve corrosion resistance. Manganese itself does not improve the corrosion resistance, because it has a higher potential (see Fig. 4). However, it reduces the harmful effect of impurities. For example, manganese increased the Ni tolerance limit [17]. The Fe/Mn ratio seems to control the influence of iron upon the corrosion rate rather than the overall Fe content [17]. Fe tolerance limit was equal to 3.2% of the Mn content regardless of the melt temperature at the time of casting and the type of casting produced, as shown in Fig. 7.

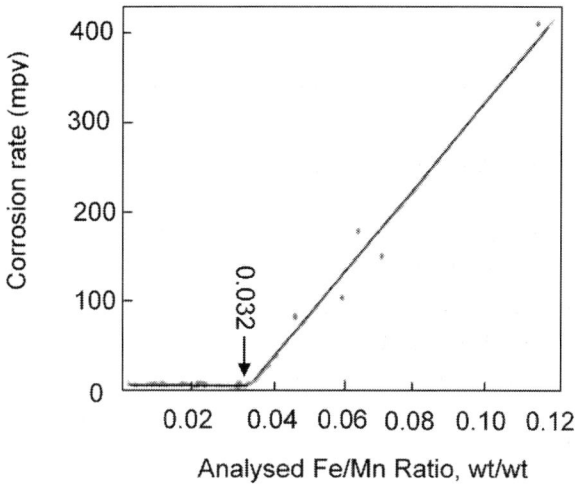

Figure 7: Effect of Fe/Mn ratio on corrosion rate of magnesium, 1 mpy≈ 40 μm/year [17].

Mn reduces the corrosion rate, probably by the following mechanisms [5]. Mn can reduce the corrosive effect of iron by wrapping the iron particles. A particle of iron embedded in a particle of manganese is less detrimental to magnesium because the galvanic activity between Mn and Mg is less than that between Mg and Fe [5]. Mn combines with the Fe and precipitates at the bottom of the crucible, where it reacts with the Fe left in suspension during solidification [19].

When Mn and Al are present together, the AlMnFe intermetallic compounds form preferentially relative to $FeAl_3$, which is known to be an active cathodic phase relative to the magnesium matrix [19]. Mn in excess of that needed to render to the Fe content ineffective could be detrimental to corrosion resistance. Low Mn binary AlMn particles: Al_4Mn, Al_6Mn, show a decreasing cathodic current density, while high Mn content Al_8Mn_5 particles show a continuously high cathodic current density. Thus, additions of Mn beyond that needed for the specified Fe/Mn ratio should be avoided [19].

Important Alloying Elements

Alloying elements not only enhance the mechanical properties of Mg but also have a significant impact on the corrosion behaviour of Mg-Al alloys. Alloying elements can form secondary particles which are noble to the Mg matrix thereby facilitating corrosion or enrich the corrosion product thereby possibly inhibiting corrosion [19].

Aluminium

Alloying magnesium with aluminium in general improves the corrosion resistance. Lunder et al reported [20] that there is a significant drop in the corrosion rate as the aluminium content is increased from 2 to 4%. Further aluminium additions up to 9% give only a modest further improvement, as shown in Fig. 8.

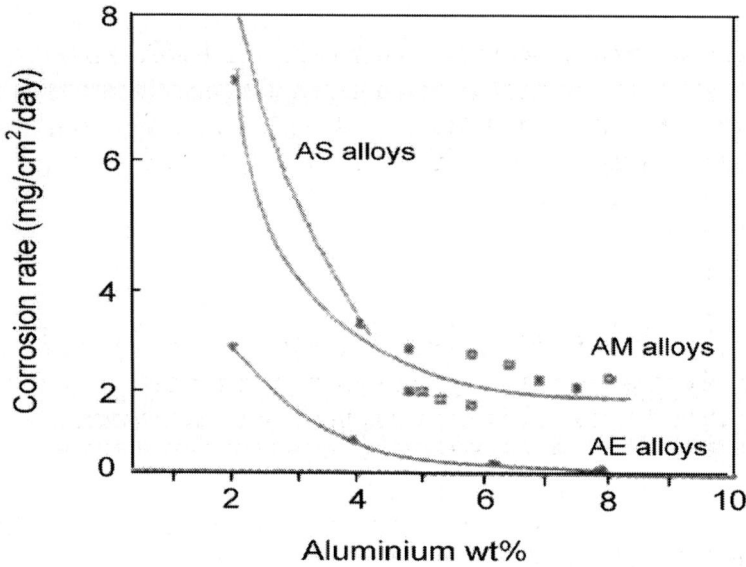

Figure 8: Corrosion rates of die ast Mg alloy immersed in 5% NaCl solution as a function of their Al content [19].

The aluminium is partly in solid solution, and partly precipitated in the form of $Mg_{17}Al_{12}$. Lefebvre and Nussbaum [21] discussed the role of Al in solid solution and role of $Mg_{17}Al_{12}$ in the corrosion process. Both Al in solid solution and $Mg_{17}Al_{12}$ can decrease corrosion rates. The presence of Al in solid solution in the matrix decreases the corrosion rate of Mg alloys in 5% NaCl and $Mg(OH)_2$ solution, which is attributed to a change in the surface microstructures.

However, it has also been found that aluminum can have a negative influence on corrosion. Aluminium reduces the iron tolerance limit from 170 wt.-ppm to 20 wt.-ppm [17]. The tolerance limit of iron decreases almost linearly with increasing aluminium content. This trend in the iron tolerance limit appears to be consistent with the formation of a passive AlMnFe intermetallic phase on solidification [13].

Zinc

The presence of Zinc can increase the tolerance limits and reduce the effect of impurities once the tolerance limit has been exceeded. Zinc is believed to improve the tolerance of Mg-Al alloys for all three impurities (Fe, Cu, Ni), but its amount is limited to 1-3%. The addition of 3% Zn raises the tolerance limit to 30 wt.-ppm Fe and greatly reduces the corrosion rate for iron concentrations of up to 180 wt.-ppm for Mg-Al-Mn alloys. For the Mg-Al-Mn–Ni alloys, 3% Zn shifts the tolerance limit from 10 to 20 wt.-ppm Ni and reduces the corrosion rate at higher concentration of nickel [13].

Zirconium

Besides the improvement of mechanical properties, the zirconium containing magnesium alloys usually have a higher corrosion resistance than zirconium-free magnesium alloys. Zr reduces the corrosion rate, probably by the following mechanisms [22]: First, impurities combine with zirconium and form insoluble precipitates and purify the alloy. The second mechanism is that zirconium stablizes the magnesium solid solution making it less soluble in aqueous solutions. The third mechanism is that zirconium combines with some intermetallic precipitates which were originally active cathodic sites, making them less active. Apart from the above mechanisms, the grain refining effect provided by zirconium could be another reason for the higher corrosion resistance of magnesium containing zirconium.

Silicon

Si is intentionally added only to the Mg-Al-Si alloys (AS series) to combine with Mg and form Mg_2Si which precipitation strengthens the alloy and is relatively innocuous to the corrosion of Mg. Mg_2Si has a steady state corrosion potential of –1.65 V similar to –1.66 V for pure magnesium in 5% NaCl solution [19].

Rare Earth Additions

It is well known that the addition of rare earth elements (RE) is an effective way to improve the mechanical properties of magnesium alloys at elevated temperatures. The improvement has mainly been attributed to the formation of a metastable RE-containing phase along the grain boundaries which significantly increases the creep resistance [23]. The beneficial effect of the RE could be similar in nature to that of the Mn additions i.e. the formation of an AlFeMn intermetallic phase which mitigates the harmful effects of Fe. However, no ternary phase was found to form in Mg-Al-RE alloys as the chemical stability of the Al_4RE intermetallic is quite high. Rather, it appeared to be more likely that the rare earths influenced the corrosion products, thereby affecting the corrosion behaviour of the Mg-Al alloys [19].

ROLE OF Β *PHASE*

Phase contents have a pronounced influence on the corrosion of magnesium, because most elements only affect the corrosion resistance of magnesium alloys when they form second phases. For example, most impurities or alloying elements form second phases and either serve as effective cathodes during the corrosion processes, or eliminate the deleterious effect of impurities, as discussed in sections 3.2.2 and 3.2.1. In this section, the role of β phase ($Mg_{17}Al_{12}$) is discussed.

The detailed microstructure is determined by the casting method, particularly the rate of solidification, and any subsequent heat treatment. Fig. 9 shows a typical microstructure of die cast AZ91D [24]. AZ91 alloy contains three main phases: a substitutional solid solution of aluminium in magnesium (α phase) and two intermetallic phases: β ($Mg_{17}Al_{12}$) present in a eutectic phase (α+β) and an intermetallic phase containing Mn, Fe and Al present at a minor level [25].

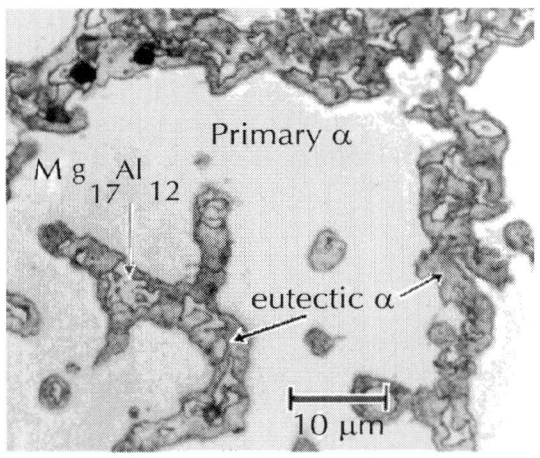

Figure 9: Microstructure of die cast AZ91D alloy [24].

In general, the β phase present in the alloy is considered more resistant to corrosion than the surrounding matrix alloy. Mathieu et al [25] studied the corrosion resistance of the different constituents of an AZ91 alloy in ASTM D1384 water, pH 8.3. It was found that Al contributed to the enhancement of the protection of the α phase through a superficial layer of a carbonate hydroxide of magnesium and aluminium. Song et al. [26] suggested that the role of the β phase in corrosion may be twofold: (1) The β-phase could have a barrier influence because it is very stable in a NaCl solution; (2) The β-phase is an efficient cathode for hydrogen evolution. First, when the β-phase is present in an alloy, it reduces the reactive surface area, so less area of the alloy is available to be corroded. Moreover, when corrosion is developing, a continuous β-phase along the α-grain boundaries might be able to prevent corrosion from spreading from one α-grain to another α-grain directly across the β-phase. Then corrosion might be stopped after the top layer α-grains have been dissolved and a continuous β-phase is exposed to solution. In this manner, the β-phase might improve the corrosion resistance of the alloy. Its barrier role is however limited to stopping the corrosion from either spreading laterally or from progressing deeper into the α-matrix. Unfortunately, in most cases the β-phase is discontinuous and then it does not stop corrosion completely. Second, the cathodic reaction of hydrogen evolution on the β-phase is much easier than on the α-phase. Consequently, if the β-phase is present in a alloy, it acts as an effective cathodic phase to the α-matrix. It causes significant acceleration of the α-phase corrosion by galvanic coupling. However, the role of β-phase strongly depends on the volume fraction. If the β-phase is present in the α-matrix as intergranular precipitates with a small volume fraction, then the β-phase mainly acts as a galvanic cathode, and accelerates the corrosion of the α matrix. If the β fraction is high, then the β-phase mainly acts as an anodic barrier to inhibit the overall corrosion performance.

MICROSTRUCTURE

Microstructural parameters such as composition, porosity, grain size, and amount and distribution of β-phase also play a role in determining the corrosion behaviour. For example, reduction of grain size increases the overall grain boundary area thereby optimizing the distribution and minimizing the size of any possible detrimental intermetallics such as $FeAl_3$. The microstructure can be controlled by the cooling rate, with more rapid cooling leading to a smaller grain size, more β-phase and a more finely distributed β-phase. For example, Mathieu et al [27] found that the casting method strongly influences the corrosion performance through control of the microstructure. The corrosion resistance of semi-solid cast AZ91D alloy is 35% higher than that of the same alloy processed by high pressure die-casting with the impurity level (Cu, Fe) being the same, and the difference in corrosion behaviour is attributed to the distribution, composition and volume fraction of the constituent phases (mainly α and β). Even for the same material, the corrosion resistance is different depending on the location of the material in the casting. For example, Song et al [24] found that the skin of die cast AZ91D shows a corrosion resistance significantly better (by nearly a factor of 10) than its interior in 1 N NaCl at pH 11. This is attributed to a higher β-fraction, and more continuous β-phase around finer α-grains and low porosity. The corrosion behaviour of the α-phases in AZ91 depends on both their aluminium contents and the local current density (which in turn depends on the details of the microstructure and the details of the environmental conditions). For high current densities, the eutectic-α at the grain boundaries tends to be corroded first. For low current densities, the primary-α, in the grain interior, would be preferentially corroded.

FORMS OF CORROSION SUFFERED BY MAGNESIUM ALLOYS

Galvanic Corrosion

When two dissimilar metals are placed in contact in a corrosive or conductive solution, a potential difference produces electron flow between them. The more active metal then becomes anodic and is corroded, and the less active metal becomes cathodic and is protected. This kind of corrosion is called galvanic corrosion, or two-metal corrosion. Magnesium and its alloys are highly susceptible to galvanic corrosion, because magnesium has the lowest standard potential of all the engineering metals as illustrated in Table 2 [288].

Galvanic corrosion can also occur between two different phases. Fig. 10 illustrated those kinds of galvanic corrosion, external and internal [13]. When magnesium and its alloys are placed contact with other metals, magnesium and magnesium alloys are corroded, while hydrogen gas is evolved on the other metals. When magnesium and magnesium alloys contain second phases because of impurities or alloying elements, the matrix α–phase is corroded, while the hydrogen gas is evolved on the second phases. Table 3 shows typical corrosion potential values for magnesium and common magnesium alloy second phases [29].

Table 2: Standard EMF series of metals [28]

	Metal – metal ion equilibrium (unit activity)	Electrode potential vs normal hydrogen electrode 25 °C, volts
↑	Au – Au+3	1.498
	Pt – Pt +2	1.2
	Pd – Pd+2	0.987
	Ag – Ag+	0.799
Noble or cathodic	Hg – Hg$_2$2+	0.788
	Cu – Cu +2	0.337
	H$_2$ – H +	0.000

	Pb – Pb^{+2}– 0.126	
	Sn – Sn^{+2}– 0.136	
	Ni – Ni^{+2}– 0.250	
	Co – Co^{+2}– 0.277	
	Cd – Cd^{+2}– 0.403	
Active or anodic	Fe – Fe^{+2}– 0.440	
	Cr – Cr^{+3}– 0.744	
	Zn – Zn^{+2}– 0.763	
	Al – Al^{+3}– 1.662	
	Mg – Mg^{+2}– 2.363	
	Na – N^{a}+	– 2.714
	K – ^{K}+	–2.925

From Table 3, it was found that the most potent cathodes in Mg-Al alloy are iron-rich phases, in particular the iron-aluminum intermetallic phase FeAl$_3$. FeAl$_3$ is one of the most detrimental cathodic phases present in Mg-Al alloys on the basis of its potential and its low hydrogen overvoltage. Al-Mn phases are also detrimental, while Mg$_2$Si seems to have no influence.

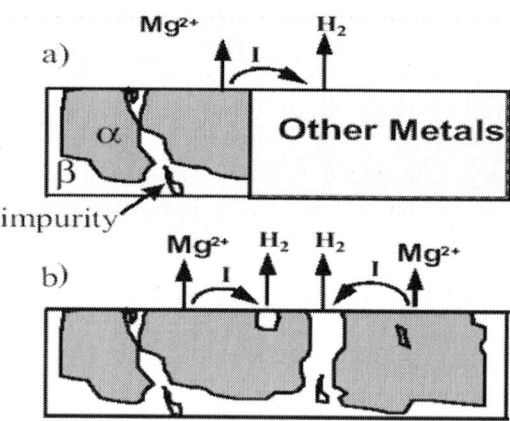

Figure 10: a) External galvanic corrosion. b) Internal galvanic corrosion [13].

Table 3: Typical corrosion potential values for magnesium and for common magnesium second phases (after 2 h in deaerated 5% NaCl solution saturated with $Mg(OH)_2$ (pH 10.5)) [29]

Metal	$E_{corr\ VSC}E$
Mg	−1.65
Mg_2Si	−1.65
Al_6Mn	−1.52
Al_4Mn	−1.45
Al_8Mn_5− 1.25 $Mg_{17}Al_{12}$− 1.20	
$Al_8Mn_5(Fe)$	−1.20
Beta−Mn	−1.17
Al_4Mn	−1.15
$Al_6Mn(Fe)$	−1.10
$Al_6(MnFe)$	−1.00
$Al_3Fe(Mn)$	−0.95
Al_3Fe	−0.74

The matrix α-phase in Mg alloys is normally anodic to the second phases and is usually preferentially corroded. Song and coworkers [24, 26] have suggested that the primary α and eutectic α phases, which have different aluminum contents, have different electrochemical behaviour. Both the primary and eutectic α can form galvanic corrosion cells with the β phase.

The galvanic corrosion rate is increased by the following factors: high conductivity of the medium, large potential difference between anode and cathode, large area ratio of cathode to anode, and small distance from anode to cathode [288]. Song et al [30] investigated the corrosion behaviour of AZ91D when it is in contact with 380 aluminium alloy, 4150 high strength steel and pure zinc. It was found that even though the galvanic effect offers some degree of cathodic protection for aluminium and zinc cathodes, the dissolution of these metals in the salt solution is still unavoidable, particularly in the region far away from the anode/cathode junction. The dissolved Zn^{2+}or Al^{3+}ions flushed to the surface of the AZ91D anode could react to form zinc or aluminium oxides or hydroxides and finally deposit on the AZ91D surface. These products can provide a certain degree of protection for the AZ91D surface.

Stress Corrosion Cracking (SCC)

Stress-corrosion cracking refers to cracking caused by the simultaneously presence of tensile stress and a specific corrosive medium [288]. Pure magnesium can be considered immune to stress corrosion cracking in both atmospheric and aqueous environments, with no reported failures occurring when loaded to its yield strength [7]. Aluminum containing alloys of magnesium are generally considered the most susceptible to SCC, with the tendency increasing with the aluminum content [7]. The alloys AZ61, AZ80, and AZ91 with 6, 8, and 9% aluminum, respectively, can show high susceptibility to SCC in laboratory and atmospheric exposures, while AZ31, a 3% aluminum alloy used in wrought product applications, is considered to show good corrosion resistance [7]. Magnesium-zinc alloys such as ZK60 and ZE41 that are alloyed with zirconium, or zirconium and rare earth elements, are typically considered only mildly susceptible, while magnesium alloys containing no aluminum or zinc are the most SCC-resistant. For example, M1 alloy, a 1% manganese alloy, like unalloyed Mg itself, shows no evidence of SCC when placed under tensile stresses as high as its yield strength [13].

SCC in magnesium is mainly transgranular. Sometimes intergranular SCC occurs as a result of $Mg_{17}Al_{12}$ precipitation along grain boundaries in Mg-Al-Zn alloys [13].

Corrosion Fatigue

There is very little research on the corrosion fatigue of magnesium alloys. It has been indicated in reference 7 that corrosion fatigue has a close relationship with humidity. For example, AZ31 subjected to an axial load cycle at 10^5 cycles per hour in air and then subjected to increasing levels of humidity showed a slow decrease in the fatigue strength once the humidity exceeded 50%. At 93% relative humidity, the measured fatigue strength had declined to about 75% of that in dry air. It has also been found [13] that corrosion fatigue cracks propagate in a mixed transgranular-intergranular mode and that the corrosion fatigue crack growth rate was accelerated by the same environments that accelerate stress corrosion crack growth. And the corrosion fatigue resistance of AZ91-T6 was significantly reduced in 3.5% salt water relative to that in air.

Pitting Corrosion

Few studies have addressed these forms of localised attack of Mg and Mg alloys because other forms of corrosion such as general, galvanic, or stress corrosion have been the cause of more serious failure of these materials. The studies of pitting of Mg and Mg alloys have been concerned with comparing the pitting behaviour of cast to that of rapidly solidified Mg alloys. Makar and Kruger [31] showed that rapidly solidified AZ61 exhibited better resistance to pitting than cast AZ61 in a buffered carbonate solution containing various levels of Cl$^-$. Pit initiation of rapidly solidified AZ61 is found to take place at a higher potential and the pit growth rate was apparently lower than cast AZ61. In a review given by Makar and Kruger [5], it is reviewed that the difference between rapidly solidified and cast Mg. A metallic glass $Mg_{70}Zn_{30}$ exhibited a better resistance to pitting. Also, the film on the metallic glass was more protective against pitting attack than the pure Mg. The glassy Mg alloy is found to exhibit a more stable passive film than pure Mg, Zn or several other crystalline Mg-based alloys. Heavy metal contamination promotes general pitting attack. In Mg-Al alloys, pits are often formed due to selective attack along $Mg_{17}Al_{12}$ network that is followed by the undercutting and falling out of grains [13].

Crevice corrosion does not occur with the Mg alloys because corrosion is relatively insensitive to oxygen concentration difference [3, 6].

Filiform Corrosion

Filiform corrosion is caused by an active corrosion cell which moves across a metal surface. The head is the anode and the tail the cathode.

Filiform corrosion occurs under protective coatings and anodized layers. Uncoated pure magnesium does not undergo filiform corrosion. However, filiform corrosion can occur on uncoated AZ91 and this indicates that a relatively resistant oxide film can naturally be formed on this alloy [13].

Oxidation at Elevated Temperatures

The research on oxidation at elevated temperatures is very limited. Magnesium corrosion resistance is typically considered to be good in dry air to about 400°C and to about 350°C in moist air [7]. At elevated temperatures, magnesium oxidizes easily in air according to a linear, kinetic oxidation curve (see Fig. 11) [6], which demonstrates the insufficient protective character of the magnesium oxides. Fig. 12 shows the effects of various elements on the oxidation rate of magnesium alloys in air at 475 °C. Additions of Pb, In, Ag, Cd, or Ti have little effect on the rate of oxidation. The additions of rare earths (Ce+La) somewhat retard the oxidation process of magnesium in air on heating. Additions of Cu, Ni, Ca, An, Sn, or even Al accelerate the oxidation rate of magnesium alloys in the air at elevated temperatures [6].

Czerwinski investigated the early stage oxidation and evaporation of AZ91D alloy [32].Thermogravimetric technique was used to determine the oxidation and evaporation behaviour of AZ91D magnesium alloys with 5 and 10 ppm of beryllium at temperatures between 200 and 500 °C. The oxidation mechanism of AZ91D changes from protective to non-protective with linear or accelerated kinetics depending on the temperature and time of exposures. During reaction in an oxidizing environment, the beryllium addition delays the onset of non-protective oxidation. It does not affect, however, the oxidation kinetics during the initial stages. During reaction in an inert atmosphere of argon, beryllium suppresses the magnesium evaporation [32]. Fig. 12 shows the effects of various alloying elements on the oxidation rate of Mg alloys in air at 475 °C.

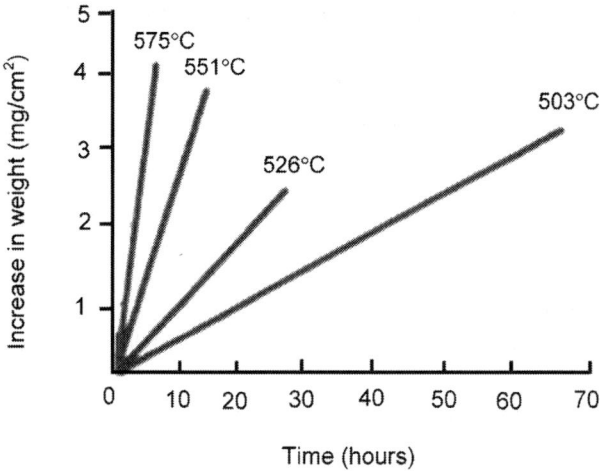

Figure 11: Pure Mg oxidation in an oxygen atmosphere under one atmospheric pressure at various temperatures [6].

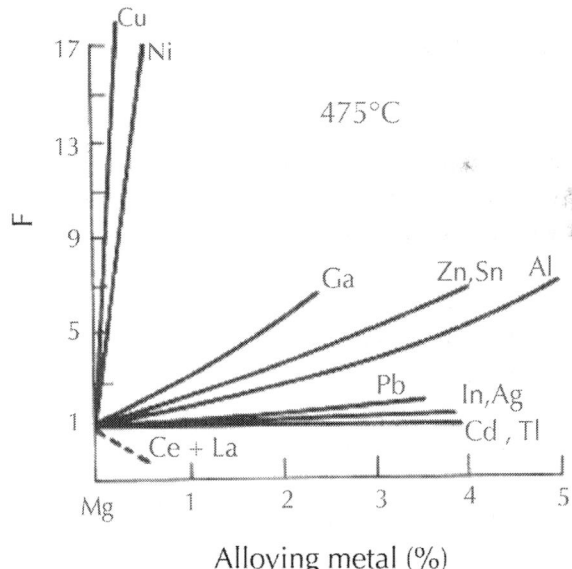

Figure 12: Effect of various alloying elements on the oxidation rate of magnesium alloys in air at 475 °C [6]. F − ratio of oxidation rate of the alloy to oxidation rate of pure magnesium.

CORROSION PREVENTION

There are a number of approaches to overcoming the corrosion problems of Mg alloys [5]: (1) High purity or new alloys: Decrease impurities to below their tolerance limits and develop new alloys with new elements, phases, and microstructure distributions; (2) Surface modification: this includes ion implantation and laser annealing; (3) Refinement of the grain size and intermetallic particles: corrosion resistance can be affected through the microstructure; and (4) Protective films and coatings.

High Purity or New Alloys

Improving corrosion resistance by producing Mg alloys with low concentration of deleterious elements is an often-used strategy. This ensures the highest possible degree of uniform corrosion resistance of the starting material [5].

Surface Modification

There are two main surface modification techniques which are discussed below.

Ion implantation is the technique whereby almost any elemental ions may be implanted into the surface of any solid using a beam of energetic ions accelerated into a target under vacuum conditions. This homogenization is the primary benefit of ion implantation in terms of corrosion resistance. Additional benefits include the ability to alter the surface while retaining the bulk properties, the creation of novel surface alloys, and the elimination of surface adhesion problems associated with coatings. The primary disadvantages are that it is a line–of–sight technique and it modifies only a thin film [33]. Akvipat and co-workers [34] examined the effects of iron implanted Mg and AZ91C in boric acid and borate buffer solution with 1000 ppm NaCl. It was known that iron degrades the corrosion resistance of magnesium alloys when introduced during conventional processing, and the goal of their work was to evaluate the effects of iron introduced by implantation. The implanted iron changed the nature of the attack on the AZ91. In the un-implanted case, $Mg_{17}Al_{12}$ islands acted as local cathodes, causing accelerated corrosion of the surrounding matrix to form a deep channel around these islands. The implanted iron shifted the attack to the $Mg_{17}Al_{12}$ particles themselves, which resulted in a more uniform attack without the rapid channelling suffered by the magnesium matrix in the un-implanted case. The results of these ion implantation studies are encouraging, but improvements in the economics and versatility of the implantation process are necessary for increasing the practical importance of this approach.

Laser Annealing

Laser annealing technique involves the formation of metastable solid solutions as promoted at metal surfaces by laser annealing, where cooling rates as high as 1010 K/s are achievable using lasers pulsed in the nanosecond range [33]. It is, therefore, another form of rapid solidification processing, but involves the melting and solidification of surface layers only. Besides the advantages of ion implantation, the advantages include the ability of lasers to treat more complex geometries, the greater depth of treatment, inexpensive operation costs, and greater control of the concentration of the modified layer [33]. The main disadvantage is the extra machining necessary because of dimensional changes during treatment. Akvipat and co-workers [34] examined the effects of thin layers of about 100nm of Al, Cr, Cu, Fe, and Ni on the pitting resistance of AZ91C in boric acid-borate solution with 1000 ppm NaCl. The role-played by these elements after laser treatment is certainly different from that when they are present in conventional processing, especially Cu, Fe, and Ni, which are detrimental, even in small concentrations under equilibrium conditions. This improvement is probably related to the structure and composition of the near-surface region [34].

Microstructure Refinement

Corrosion resistance can be affected through modification of the microstructure. Recent studies [19, 35,36] have centered on the refinement of the grain size and intermetallic particles and on developing a more homogeneous microstructure.

Grain Refinement

Reduction of grain size increases the overall grain boundary area thereby optimizing the distribution of detrimental intermetallics and minimizing the size of any possible detrimental intermetallics such as $FeAl_3$. The traditional grain refinement method in sand casting is to add an innoculent which facilitates heterogeneous nucleation during solidification. Indeed, additions of strontium to Mg-Al alloys have shown a marked reduction in grain size but also have pointed to a possible alteration of both the oxide layer structure and composition and the electrochemical properties of the phases present [19].

Effect of Rapid Solidification Processing

Rapidly solidified materials show improved corrosion resistance because of a refined microstructure which translates to a more homogeneous composition thereby minimizing the potential of any microgalvanic corrosion cell.

Govind et al [35] claimed that technology of making rapid solidification ribbon of highly reactive Mg-9%Al-1%Zn-0.2%Mn alloy was established successfully. Grain sizes of 1-3 μm could be achieved in as-spun ribbons in contrast to a 250-300 μm grain size normally attained in sand cast structure. Below a temperature of 200 °C no grain growth was observed in RS ribbons of Mg alloy as the precipitates of intermetallic compound $Mg_{17}Al_{12}$ pin the grain boundaries.

Effect of Heat-Treatment

Heat-treatment can drastically alter the size, amount and distribution of the precipitated   phase, $Mg_{17}Al_{12}$, which in turn, alters the corrosion behaviour of Mg-Al alloys. Aung and Zhou [36] studied an AZ91D ingot in the as-cast condition that was homogenized by solution treatment and then aged for various periods of time. The homogenization treatment of an AZ91D ingot at 420 °C for 24h was found to be effective in dissolving the -precipitates. Artificial ageing at 200°C caused precipitation of the -phase mainly along the grain boundaries. The volume fraction of the -phase was observed to increase with ageing time. A homogenization treatment improved the corrosion resistance of the AZ91D ingot, but ageing for 8 h, 16 h or 26 h lowered the corrosion resistance. These results support the suggestion that microgalvanic coupling exists between the cathodicβ-phase and the anodic α-matrix. The inhibiting effect of the β-phase in the artificially aged alloy predominated during the short interval of electrochemical testing but the accelerating effect of the decrease in aluminium content in the matrix predominated in the long period immersion testing. During immersion testing, the phase may dissolve into the chemical solution and this also tends to accelerate the corrosion rate.

Protective Coatings and Films

There are a number of technologies available for coating magnesium and its alloys. These include electrochemical plating, conversion coatings, anodizing, hydride coatings, organic coatings and vapour-phase processes.

Electrochemical Plating

One of the most cost effective and simple techniques for introducing a metallic coating to a substrate is by electrochemical plating. The electrochemical plating can be subdivided into two types: electroplating and electroless plating. In both cases a metal salt in solution is reduced to its metallic form on the surface of the substrates. In electroplating the electrons for reduction are supplied from an external source. In electroless or chemical plating the reducing electrons are supplied by a chemical reducing agent in solution or, in the case of immersion plating, the substrate itself.

Electroplating:

Besides some traditional disadvantages of electroplating such as non-uniform coatings and difficulties in coating complex shapes, there are some challenges to be met for electroplating on magnesium. The pre-treatment processes are complicated due to the fact that, in the presence of air, magnesium very quickly forms a passive oxide layer. Cu-Ni-Cr plating has been shown to have good corrosion resistance in interior and mild exterior environments [37]. Also, it is necessary to develop non-traditional plating baths since magnesium reacts violently with most acids and dissolves in acidic media. Furthermore, magnesium and its alloys are prone to galvanic corrosion, so the metal coating must be pore free otherwise the corrosion rate will increase. Usually, the coating is at least 40-50 μm thick to ensure pore-free coatings. Furthermore, the alloys are difficult to plate because intermetallic species such as Mg_xAl_y are formed at the grain boundaries, resulting in a non-uniform surface potential across the substrate, and therefore further complicating the plating process. Recently, Jiang et al [388] studied Zn-Ni alloy coatings pulsed-plated on magnesium alloy AZ91. Before deposition, the substrate surfaces were processed in a standard industrial way: polishing with alumina sand paper, alkaline degreasing, chemical pickling, activation, zinc immersion and Zn-Cu alloy plating. A Zn layer and a Zn-Cu layer under the Zn-Ni coating were applied to improve the adhesion and to protect the substrate using the small electrode potential difference between Zn-Cu and Zn-Ni layer. Zn-Ni coatings were deposited in an alkaline bath with a composition as ZnO 10g/l, NaOH 150 g/l, $NiSO_4 \cdot 6H_2O$ g/l, triethanolamine 50 g/l, and at 10-40°C, 500-4000 Hz, 0.04-0.1 A/cm². The bonding strength can be as high as 14.8 MPa. The corrosion life of the Zn-Ni coating can reach over 200 h in a salt spray test conducted according to ASTM B1117. However, no detailed data were given on the porosity of the coatings, which may increase the corrosion rate due to the galvanic corrosion effect.

Electroless plating:

Electroless plating has good throwing power and can produce a uniform coating thickness on complex objects. It also involves a simple pre-treatment and is suitable for magnesium alloys with high aluminium contents [1]. However, electroless deposited coatings cannot be too thick, the bath life is limited, and deposition rates are slow. In particular, electroless plating requires the use of hydrofluoric acid during the pre-treatment, which increases the danger of the operation and is not environmentally friendly [1]. Research on increasing the bath life and eliminating toxic chemicals is necessary in order to create a green-plating process for coating magnesium. Sharma et al. [39] studied the properties of an electroless nickel coating on magnesium alloy ZM21. The solution contains nickel carbonate, sodium hypophosphite (metal-reducing agent), citric acid and bifluoride (act as accelerators, complexing agent, and accelerators), thiourea (solution stabilizer and brightening agent) and ammonia solution. The paper put forward some reactions, and suggested that the autocatalytic reaction for nickel deposition is initiated by catalytic dehydrogenation of the reducing agent with release of hydride ion, which then supplies electrons for the reduction of nickel ions.

The coated samples were immersed in a 5% solution of sodium chloride at pH 7.0. No corrosion spots on the coatings were noticed after 96 hrs of immersion. The formation of corrosion spots initiated only after the fifth day of immersion. Recently, Huo et al.[40] developed an environmental-friendly combined technique of chemical conversion treatment followed by electroless nickel plating for AZ91D alloy to improve corrosion resistance. The presence of the conversion coating, which was mainly $MgSnO_3 \bullet H_2O$, between the nickel coating and the substrate reduced the potential difference and avoided any catastrophic galvanic corrosion between nickel and magnesium. The electroless nickel coating containing about 10 wt% phosphorus greatly enhanced the corrosion potential of AZ91D from $-1.50\,V$ to-0.60 V.

Conversion Coatings

Conversion coatings are produced by chemical or electrochemical treatment of a metal surface to produce a superficial layer of substrate metal oxides, chromates, phosphates or other compounds that are chemically bonded to the surface. On magnesium, these coatings are typically used to enhance paint adhesion to the coatings and provide improved corrosion protection to the metal. There are a number of different types of conversion coatings including chromate, permanganate, phosphate, phosphate-permanganate and fluorozirconate treatments. The conventional conversion coatings are based on chromium compounds that have been shown to be highly toxic carcinogens. The development of an environmentally friendly process is a necessity due to the more stringent environmental protection laws currently in effect, or being proposed. The coatings on alloys also represent a significant challenge due to their non-uniform surface composition.

Phosphate–Permanganate Conversion Coatings

Phosphate-permanganate treatments are being explored as an alternative to conventional chromate conversion coatings. These treatments are more environmentally friendly and have been shown to have corrosion resistance comparable to chromate treatments [1].

Chong and Shih [41] reported that a conversion coating on magnesium alloys AZ61A, AZ80A and AZ91D prepared from a solution containing permanganate ($KMnO_4$ 20g/l) and phosphate ($MnHPO_4$ 60g/l) showed an equivalent or slightly better passive capability than a conventional chromate-based conversion treatment, but an inferior passive capability for the pure Mg specimen. Hawke and Albright [42] studied a phosphate-permanganate treatment for the conversion coating of AM60B. The coating is based on magnesium phosphate, but contains significant amounts of aluminum compounds generated from the alloy's aluminum content, and manganese compounds formed by reduction of the permanganate ion. The manganese is considered to contribute manganese to the coating, and acts as an accelerator without depositing metallic manganese on the magnesium surface. The coatings were shown to have good corrosion resistance and paint adhesion.

It was found that the most important factor in producing the best quality conversion coatings was the control of the pH [1]. Since pH is the most important factor determining conversion coating quality, the research on stabilizing the pH of solutions has gained increasing attention. Umehara et al. [43] claimed that a pH-stabilizing solution was developed for the conversion coating on AZ91D. The surface film formed was composed of magnesium oxide, and manganese oxide, and contained boron oxide. The pH change was insignificant with increasing the surface area of the magnesium treated. After cleaning and surface activation, the samples were immersed in a solution containing potassium permanganate and either nitric or hydrofluoric acid. The coatings formed in the bath containing nitric acid were substantially thicker and crystalline manganese oxide was observed. The corrosion resistance of these coatings was equivalent to the protection afforded by a standard chromate treatment.

Stannate Conversion Coatings

A study on stannate treatment of ZC 71 and a metal matrix composite of ZC71+12% SiC particles has been undertaken by Gonzalez-Nunez et al [44]. After mechanical finishing and pickling, the samples were immersed in a stannate bath for selected periods of time. The treatment resulted in the formation of a 2-3 µm thick, continuous and adherent, crystalline coating of $MgSnO_3$ on both materials. The nucleation and growth of the coating was completed in about 20 min. The initial nucleation was found to occur at cathodic sites on the surface with crystal growth to a grain size of about 2-5 µm until they coalesced. There was an increase in the corrosion potential of the magnesium surfaces as the film formation proceeded indicating that the coating does have a passivating effect on the surface.

Rare Earth Process

The corrosion protection of cerium, lanthanum and praseodymium conversion coatings on magnesium and magnesium alloy WE 43 has been investigated by Rudd et al [45]. The samples were polished, cleaned in water and methanol and dried prior to immersion in a $Ce(NO_3)_3$, $La(NO_3)_3$ or $Pr(NO_3)_3$ solution. A visible, adherent but easily removed coating was produced on the surface. It has been demonstrated that these coatings provide an increase in corrosion resistance for magnesium and its alloys. However, the coatings deteriorated on prolonged immersion in the test buffer solution so their protective effect is short term.

Conversion coatings have been known for some time, but it should be mentioned that a great deal of the work done on conversion coating of magnesium substrate is proprietary in nature. Thus, there is still a great deal of research to be done to better understand the surface reactions between magnesium based substrates and coatings [1].

Anodizing

Anodizing is an electrolytic process for producing a thick, stable oxide film on metals and alloys. These films may be used to improve paint adhesion to the metal, as a key for dying or as a passivation treatment.

The stages for processing include [1]: (1) mechanical pre-treatment, (2) degreasing, cleaning, and pickling, (3) electrobrightening or polishing, (4) anodizing using AC or DC current, (5) dying or post-treatment and (6) sealing. Sealing of the anodized film is necessary in order to achieve an abrasion and corrosion resistant film. This can be accomplished by boiling in hot water, steam treatment, dischromate sealing and lacquer sealing [1]. One of the main challenges for producing adherent, corrosion resistant, anodic coatings on magnesium results from the eletrochemical inhomogeneity due to the phase separation in the alloy. Another disadvantage of this technique is that the fatigue strength of the base metal can be affected by localized heating at the surface during the treatment.

Dow 17 process: Chemical treatment no.17, developed by Dow Chemicals, can be applied to all forms and alloys of magnesium [46]. The anodizing bath employed in this treatment is a strongly alkaline bath consisting of an alkali metal hydroxide and a fluoride or iron salt or a mixture of the two. This process produces a two-phase, two-layer coating. The first layer is deposited at a lower voltage and results in a thin, approximately 5 μm, light green coating, The over layer is formed at a higher voltage. It is a thick dark green, 30.4 μm, layer that has good abrasion resistance, paint base properties and corrosion resistance [46].

HAE process: named after its discoverer, H.A. Evangelides [47]. This treatment can be applied to all magnesium alloys including the rare-earth magnesium alloys [47]. The HAE bath is a strongly alkaline and oxidizing solution, consisting of potassium-hydroxide-aluminate-fluoride-manganate and tribaisc sodium phosphate [48]. The treatment produces a two phase coating as in the DOW 17 process [46]. At a lower voltage a 5 μm thick, light tan subcoating is produced. At a higher voltage a dark brown, thicker (30 μm) film is produced. Upon sealing the HAE treatment provides excellent corrosion resistance. The corrosion resistance of AZ91D treated with this technique has been tested by a 3-year atmospheric exposure experiment. Superior corrosion reistance compared to conversion coating was observed [49].

Other processes: Mizutani et al. [50] studied the electrochemical behaviours of pure magnesium, AZ31 and AZ91 in 1 mol/dm³ NaOH during the anodizing process. The anodizing films on Mg alloys at 3

V had the best effective corrosion resistance and these films consist of magnesium hydroxide. However, the coatings were really thin and the film thickness of anodized AZ91 at 3, 10 and 80 V was approximately 4, 1, and 0.5 μm, respectively.

Gas-Phase Deposition Processes

Protective coatings can also be produced from the gas phase. These are typically metallic coatings but can include organic coatings such as thermal spray polymer coatings and diamond like coatings. All of these processes have the advantage that they have little negative environmental impact. However, the capital costs associated with these techniques are usually high [1].

Chemical Vapour Deposition (CVD)

Chemical vapour deposition can be defined as the deposition of a solid on a heated surface via a chemical reaction from the gas phase. Advantages of this technique include deposition of refractory materials well below their melting points, achievement of near theoretical density, control over grain size, processing at atmospheric pressure and good adhesion [51]. However, CVD is limited to substrates that are thermally stable at $\geq$ 600 °C. Efforts are underway to reduce the high temperature requirements and plasma and organometallic CVD processes offset this problem somewhat. A further disadvantage of this process is high-energy cost due to the need for high deposition temperatures and sometimes low efficiency of the process.

A plasma-assisted CVD technique has been successfully used to deposit SiO_x thin films on magnesium alloy WE43 [52]. The coatings were deposited at low temperature (T< 60 °C) and 100 mTorr of pressure. Pre-treatments were performed immediately before the application of the SiO_x coating, in plasma fed with oxygen, hydrogen or CF_4–O_2 (20%). The SiOx coatings exhibited better corrosion resistance in 0.1 M NaCl with the pre-treatment in H_2 plasma than in CF_4-O_2 (20%) plasma. When the magnesium surface is treated in H_2 plasma, a preferential removal of OH groups occurs, resulting in a clean surface. The improvement of corrosion resistance of pre-treated magnesium alloy in CF_4-O_2(20%) plasma has been attributed to the formation of MgF_2 [52].

Diamond-Like Carbon Films (DLC)

Diamond like carbon films can be produced using a number of different processes such as physical vapour deposition (PVD), CVD and ion implantation. These coatings are desirable for many applications due to their high hardness, low friction coefficient, electrical insulation, thermal conductivity and inertness. Yamauchi et al. [53] reported that DLC films were deposited on the magnesium alloy substrate (Al 2.4 wt%, Zn 0.87 wt %, Si 0.001 wt%, Mn, Cu, Ni, Fe non-detected) by the plasma CVD method using radio frequency. The DLC coating was confirmed to be effective in decreasing the friction coefficient and improving the corrosion resistance in 3 wt% NaCl and 0.05 N NaOH solutions. However, DLC films showed poor corrosion wear resistance in 0.05 N HCl due to the existence of pits in the films.

Physical Vapour Deposition Processes

PVD involves the deposition of atoms or molecules from the vapour phase onto a substrate. There are a few challenges to overcome in the PVD coating of magnesium substrates. The deposition temperature must be below the temperature stability of magnesium alloys (180 °C) and good adhesion must be obtained despite this low temperature. Hollstein et al [54] compared the mechanical and chemical properties of various PVD coatings on a high purity AZ31 magnesium alloy, including single layer TiN, CrN, double layer TiAlN, NbN-(TiAl)N, CrN-TiCN, the multi-layer composite AlN/TiN, and superlattices CrN/NbN. The difference between the (TiAl)N layers and the TiN/AlN multilayers is that the (TiAl)N layers were produced using a Ti-50%−Al-50% target compound, whereas the TiN/AlN multilayers are produced by power switching between a titanium target and an aluminium target. NbN/CrN superlattices are characterised by a repeating layer structure of the two materials with a nanometric scale dimension. The best results in corrosion resistance, adhesion and hardness were obtained from the CrN and (TiAl)N coatings. The classical TiN monolayer coating with a thickness that is typical for decorative purposes (about 1 μm) is not suitable to protect Mg alloys against corrosion effectively. It seems that a minimum thickness of about 4 μm or more is necessary for industrial applications. Hoche et. al [55] developed a new method of plasma anodisation to ensure acceptable corrosion resistance, besides excellent wear protection on Mg alloy. The anodizing and PVD-coating can be done in one process. The 0.5 μm plasma anodisation layer and 1.5 μm PVD-Al_2O_3 coating was subjected to 120 hours salt spray.

Organic/Polymer Coatings

Organic finishing is typically used in the final stages of a coating process. These coatings can be applied to enhance corrosion resistance, abrasion and wear properties, or for decorative purpose. An appropriate pre-treatment process is required in order to produce coatings with superior adhesion, corrosion resistance and appearance [1]. Many coating processes can be applied to magnesium and magnesium alloys, including painting, powder coating, e-coat, sol-gel process, and polymer plating. In the following section, we discuss the sol-gel process.

Sol-Gel Process

Synthesis of gels by the sol-gel process involves the hydrolysis and condensation polymerization of metal alkoxides. One of primary advantages of this technique is the excellent adhesion obtained with a minimum of sample pre-treatment [1]. The metal surfaces are simply degreased, rinsed and dried prior to dip-coating in the sol-gel mixture. A significant advantage is that irregular shapes and larger integral structures can also be coated. However, sol-gel coatings tend to fail if the film thickness exceeds 5 μm because of shrinkage strains during drying, and densification of the as-deposited xerogel film. Phani et al [56] reported that sol-gel coatings consisting of ZrO_2 as well as 15 wt% of CeO_2 could be deposited on magnesium alloys AZ91D and AZ31 by the dip coating technique. Adhesion measurements on the coatings showed good adhesion with critical loads of up to 25 N. Depth-sensing nanoindentation tests of the coatings showed a hardness of around 4.5 GPa and an elastic modulus of 98 GPa. Coatings deposited on AZ91D and AZ31 substrates exhibited good corrosion resistance in the salt spray test performed for 96 hr.

Electrolytic Plasma Oxidation

Electrolytic Plasma oxidation (EPO), also called plasma anodising or micro arcing, is a promising surface treatment for hexavalent chromium replacement in anti-corrosion protection or in the improvement of the tribological properties of lightweight metal structures. This electrolytic plasma oxidation can be distinguished from classical anodising by the use of voltages above the dielectric breakdown potential of the anodic oxide being formed. This leads to the local formation of plasmas, as indicated by the presence of sparks that are accompanied by a release of gas [57].

Some interesting historical comments are presented in a review by Yerokhin et al. [57]. Plasma anodizing dates back at least to 1932. At that time it was studied by two German scientists, Gunterschulze and Betz, while working on electrolytic capacitors using aluminium foil. During the 1970s oxide deposition on an aluminium anode under an arc discharge condition was also developed and studied. In the 1980s, the possibilities of utilising surface discharges in oxide deposition onto various metals were studied in more detail. Early applications were introduced in the textile and aerospace industries. Electrolytic plasma oxidation (EPO) is recently considered to be a promising technique to deposit ceramic coatings on magnesium alloys for corrosion protection.

The EPO process involves anode electrochemical dissolution, the combination of metal ions with anions to form ceramic compounds, and sintering on the substrate under the action of the sparks. Yerokhin et al [58] described three main steps leading to ceramic coating formation. First, a number of discrete discharge channels are formed in the oxide layer as a result of loss in its dielectric stability in a region of low conductivity. The material in the channel is heated-up to temperatures of 10^4 K by the generated electron avalanches. Due to the strong electric field, the anionic components are drawn into the channel. Owing to the high temperature, the elements are melted out of the substrate, enter the channel and are oxidized. Second, these oxidized metals are ejected from the channels into the coating surface in contact with the electrolyte, thereby increasing the coating thickness in that location. In the last step, the discharge channels are cooled and the reaction products are deposited on to its walls. The above process repeats itself at a number of discrete locations over the entire coating surface, leading to an overall increase in the coating thickness. However, there is no support experimentation to confirm the above interpretation.

Yerokhin et. al. [57] also described current –voltage characteristics during the EPO process. Fig. 13 is the current-voltage diagram for the process of plasma electrolytic oxidation. First, the previously formed passive film begins to dissolve at point U_1 which, in practice, corresponds to the corrosion potential of the material. Then, in the region of re-passivation, U_1-U_2, a porous oxide film grows, and it is across this film that most of the voltage drops now occurs. At point U_2 the electric filed strength in the oxide film reaches a critical value beyond which the film is broken through due to impact or tunnelling ionisation. In this case, small luminescent sparks are observed to move rapidly across the surface of the oxide film, facilitating its continued growth. At point U_3, the mechanism of impact ionisation is supported by the onset of thermal ionisation processes and slower, larger arc discharges arise. In the region U_3-U_4, thermal ionisation is partially blocked by negative charge build-up in the bulk of the thickening oxide film, resulting in discharge-decay shorting of the substrate. This effect determines the relatively low power and duration of the resultant arc discharges, i.e. micro-discharges, which are termed "micro arcs". Owing to the 'micro-arcing', the film is gradually fused and alloyed with elements contained in the electrolyte. Above the U_4, the arc micro-discharges occurring throughout the film penetrate through to the substrate and transform into powerful arcs, which may cause destructive effects such as thermal cracking of the film.

Recently, Wang et al. [59] reported different characteristics of oxidized coatings at different voltages on AZ91D in an alkali-silicate solution. Three types of oxide coatings, passive film, micro-spark ceramic coating and spark ceramic coating, were prepared at 100 V, 195 V, and 235 V, respectively. The passive films are thin and cannot provide effective protection to the substrate. The micro-spark ceramic coatings are homogeneouswith compact internal layer and exhibit the highest resistance owing to to the highest effective thickness. The spark ceramic coatings are thickest, but due to the large pores in the oxide layer, they are loose and defective.

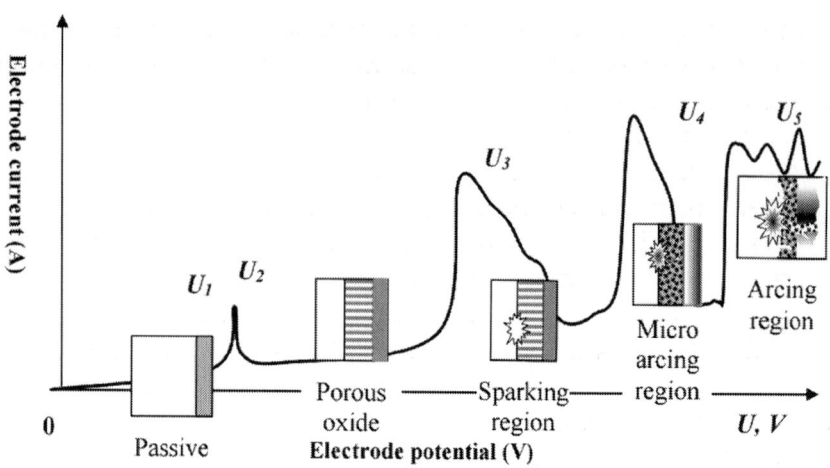

Figure 13: Current-voltage diagram for the process of plasma electrolysis: discharge phenomena are developed in the dielectric film on the electrode surface [57].

Electrolytic plasma oxidation (EPO) technology has been used for depositing ceramic coatings on magnesium alloys for corrosion protection [60-63]. The coatings can be as thick as a few hundreds of micrometers and their corrosion behaviour strongly depends on the process parameters employed, the chemical compositions of the materials studied, and the electrolytes used. The effect of electrolyte composition on properties of EPO oxide coatings on Mg and Mg alloys has also been an interesting subject of investigation to the automobile industry. The electrolytes consisted of potassium hydroxide and some other passive agents that can modify the characteristics of the oxide coatings. Hsiao and Tsai [62] studied the characteristics of anodic films formed on solutions containing 3M KOH, 0.21M $NaPO_4$, 0.6 MKF, with and without $Al(NO_3)_3$. It was found that the addition of $Al(NO_3)_3$ into 3 M KOH+0.21 M Na_3PO_4+0.6 M KF base electrolyte assisted uniform sparking on AZ91D magnesium alloy in anodizing. Either with or without a low concentration of $Al(NO_3)_3$, a porous and non-uniform anodic was formed. The presence of $Al(NO_3)_3$ in the base electrolyte resulted in the formation of Al_2O_3 and $Al(OH)_3$ in the anodic film. The presence of Al_2O_3 in the films is beneficial to the corrosion resistance of films in 3.5 wt% NaCl solution.

The process parameters employed also play an important role in the characteristics of oxide coatings. Zhang et al. [61] found that the properties of oxide coatings were strongly influenced by the process parameters employed. With an increase of solution temperature, the film thickness decreased. On the other hand, the film thickness increased with an increase in treatment time and current density. The voltages rise during the EPO process is always accompanied by the increase of film thickness. Higher voltage indicates thicker film. Khaselev et al [60] investigated the characteristics of the oxide coatings on binary Mg-Al alloys in a solution containing 3 M KOH, 0.6 M KF, and 0.21 M Na_3PO_4 with 1.1 M aluminate. The breakdown voltages increased with an increase in Al content in the alloys. The growth of oxide films was non-uniform. The growth started on -Mg and continued on the -phase $(Mg_{17}Al_{12})$ when the voltage exceeded 80 V, and a uniform anodic film was formed on the alloy substrate when the voltages reached 120 V. Al was incorporated into oxide coatings from both the substrate and the electrolyte. The EPO coatings exhibited better corrosion resistance than the coatings treated by anodizing. Zhang et al [61] compared the oxide coatings produced by the EPO process with the anodic coatings prepared by the HAE and Dow 17 process. It was found that the EPO coatings were smooth, uniform, in contrast to rough, patchy film produced by HAE and relatively rough, even, partly powdery film produced by DOW17. Furthermore, the films produced by the EPO process are much more corrosion protective than those produced by HAE and DOW17.

Now studies on the effects of power supply modes on EPO coating properties have been paid more and more attention. Researchers [64] tried to modify the morphologies and structures of oxide coatings by altering the sparks during the EPO process. Originally, DC or amplitude-modulated AC was used in the EPO process, which allowed coating growth rates of only 1-2 μm min^{-1}. Yerokhin et al [64] utilized a pulsed bipolar current to make oxide ceramic layers dense and uniform with a fine-grained microstructure on a Mg alloy (2% Al, 1% Zn, 0.2% Mn, Mg-balance) and the coating growth rates was up to 10μ m min^{-1}. The pulsed bipolar current was also beneficial to elimination of the fatigue cracks by strain distortion of the metal subsurface layers induced during the oxidation process. The phase was mainly $MgAl_2O_4$ using pulsed bipolar power, while MgO and Al_2O_3 are mainly in the films using DC power.

Other methods have also been applied to the EPO process to improve the properties of coatings. Guo et al [65] demonstrated that ultrasonic power can play an important role in the coating formation and enhance the coating growth. The anodic coatings consisted of two layers when the ultrasound field was applied and the acoustic power value increased to 400W at a constant frequency of 25 kHz in 0.1 M potassium hydroxide, 0.15 M potassium fluoride, 0.30 M sodium aluminate, 0.004 M sodium pyrophosphate and 0.5-1.0 M additives. This was different from the situation without an ultrasonic field where the anodic coatings consisted of only one layer. For the two layer anodic coatings, the inner layer was compact and enriched in aluminum and fluorine, and had a uniform thickness. In contrast, the aluminum and fluorine contents in the external layer were very low and the thickness was non-uniform. Also, studies [66-73] demonstrate that Plasma Electrolytic Oxidation (PEO) is a relatively cost-effective and environmentally friendly technique to improve the corrosion and wear resistance of magnesium and its alloys. The PEO method can be used to form a thin or thick, hard and adherent ceramic-like coating on the surface of Mg alloys for automotive applications.

It has been indicated [74-77] that magnesium is a good candidate as an implant material due to its bioabsorbability and high specific strength. To avoid the rapid degradation of the magnesium in the human body, techniques of surface treatment can be applied to improve magnesium corrosion resistance and consequently mitigate its degradation. Very recently, Hu and Nie [78] applied the plasma electrolytic oxidation (PEO) treatment to pure magnesium in an effort to develop an implant material with controllable degradation. This is because the PEO process is inexpensive and environmentally friendly and capable of producing a coating that is non-harmful to the human body. Magnesium is a strong candidate as due to its bioabsorbability and high specific strength. In their study, Potentiodynamic polarization corrosion tests, performed in a simulated body fluid (Hanks' Balanced Salt Solution) were carried out on coated and uncoated magnesium samples. The results of the testing showed that the coated magnesium exhibited higher corrosion resistance than the substrate. With the PEO coating thicknesses of 6.3 and 18.6 microns, the corrosion current density was decreased by 1.330×10^{-3} and 1.341×10^{-3} mA/cm^2 from the uncoated magnesium respectively, indicating a significant reduction in the degradation rate between pure magnesium and coated magnesium from 6.17×10^{-1} to 1.91×10^{-2} and 1.42×10^{-2} g/year respectively. A pin-on-disc tribometer was employed to measure the coefficient of friction (COF) for the coated and uncoated magnesium samples, lubricated with and without Hanks' solution. The measured COFs of the coated samples were very low. They were averaged to be 0.198 and 0.256 for the thick and thin coatings respectively, while the substrate exhibited an averaged COF of 0.203 under the lubricated condition. The COF measurements indicated the coatings had very comparable COFs to the substrate. By maintaining the low level of the COFs, the developed PEO coating on the Mg substrate could cause almost no irritation or harm to the surrounding tissue during the implant insertion operation.

SUMMARY OF THE LITERATURE REVIEW

The examples described in the previous section demonstrate that it is possible to develop appropriate coating schemes for the protection of magnesium for use in automotive components. However, no single coating technology has been developed which functions to adequately protect magnesium from corrosion in harsh service conditions. For example, in winter, the mixture of deicing salt and sand often attacks Mg chassis components of automobiles, such as transfer cases. Also, coolants can cause corrosion on Mg alloy engine blocks.

One of the most effective ways to prevent corrosion is to coat the base material. In order for a coating to provide adequate corrosion protection for Mg and Mg alloys, the coating must be uniform, pore free, well adhered, and self-healing in case that physical damage to the coating may occur.

In the case of electrochemical plating, the capital investment is relatively small. Electroplating process is extremely difficult to achieve uniform coatings on complex shapes due to uneven throwing power. Electroless plating can obtain uniform coatings. However, there are some serious concerns over waste disposal.

Conversion coatings also represent a minimum capital investment, however the most widely used type of conversion coatings are chromate conversion coatings. These represent a serious environmental risk due to the presence of leachable hexavalent chromium in the coatings. A number of chromate free conversion coatings are under development but this technology is still in its infancy. Conversion coatings do not provide adequate corrosion and wear protection from harsh service conditions when used alone. However, they can act as a good base for producing adherent organic coatings and act to enhance corrosion resistance of a combined coating system by protecting the substrate at the defect sites in overlying layers.

Anodizing is the most widely commercially used coating technology for magnesium and its alloys. This process is technologically more complex than electroplating or conversion coating but is less sensitive to the type of alloy being coated. It does involve more capital investment due to the need for cooling systems and high power consumption but this may be balanced by the decreased cost of waste disposal. The coatings produced by anodizing are porous ceramic-like coatings. These properties impart good paint-adhesion characteristics and excellent wear and abrasion resistance to the coating. However, without further sealing, they are not adequate for use in applications where corrosion resistance is of primary importance.

The use of gas-phase coating processes and laser surface melting to modify the surface or create coatings is an excellent alternative for corrosion protection of Mg alloys with little environmental impact. However, the capital investment of equipment is really high.

Organic coatings such as sol-gel coatings on magnesium alloys require a minimum of pre-treatment steps prior to deposition. But, their thicknesses might be limited.

It is evident that there is limited information comparing erosion and corrosion properties of the ceramic coatings prepared from different electrolytes. The relative significance of the EPO process parameters on effects of the corrosion resistance of magnesium alloys, in particular for bio-applications, needs to be fully investigated.

REFERENCES

1.   J. E. Gray and B. Luan, "Protective coatings on magnesium and its alloys – a critical review", Journal of Alloys and Compounds, 336 (2002) 88-113.

2.   D. Eliezer, E. Aghion and F. Froes, "Magnesium science, technology and applications", Advanced Performance Materials, 5 (1998) 201-212.

3.   W. A Ferrando, "Review of Corrosion and Corrosion Control of Magnesium Alloys and Composites", Journal of Materials Engineering, 11 (1989) 299-313.

4.   R. Lindstom, L. Johansson, G. Thompson, P. Skeldon and J. Svensson, "Corrosion of magnesium in humid air", Corrosion Science, 46 (2004) 1141-1158.

5.   G. L. Makar and J. Kruger, "Corrosion of magnesium", International Materials Reviews, 38 (1993) 138-153.

6.   N. D. Tomashov, "Theory of corrosion and protection of metals", The Macmillan Company, New York, 1966.

7.   R. Baboian, S. Dean, H. Hack, G. Haynes, J. Scully, D. Sprowls, "Corrosion Tests and Standards: Application and Interpretation", ASTM Manual Series: MNL 20, Philadelphia, USA, 1995.

8.   G. Song, A. Atrens, D. StJohn, J. Nairn and Y. Li, "The electrochemical corrosion of pure magnesium in 1N NaCl", Corrosion Science, 39 (2004) 855-875.

9.   Metals Handbook, Corrosion 9th ed., Vol. 13, "Corrosion of magnesium and magnesium alloys", ASM International, Metals Park, OHIO, USA, 1987.

10. G. Song, A. Atrens, D. St John, X. Wu and J. Nairn, "The anodic dissolution of magnesium in chloride and sulphate solutions", Corrosion Science, 39 (1997) 1981-2004.

11. G. Song and D. StJohn, "Corrosion behaviour of magnesium in ethylene glycol", Corrosion Science, 46 (2004) 1381-1399.

12. N. S. McIntyre and C. Chen, "Role of impurities on Mg surfaces under ambient exposure conditions", Corrosion Science, 40 (1998) 1697-1709.

13. G. Song and A. Atrens, "Corrosion Mechanisms of Magnesium Alloys", Advanced Engineering Materials, 1 (1999) 11-33.

14. R. Lindstom, L. Johansson and J. Svensson, "The influence of NaCl and CO2 on the atmospheric corrosion of magnesium alloy AZ91", Materials and Corrosion, 54 (2003) 587-594.

15. G. Song and D. StJohn, "Corrosion of magnesium alloys in commercial engine coolants", Materials and Corrosion, 56 (2005)15-23.

16. E. Slavcheva, G. Petkova and P. Andreev, "Inhibition of corrosion of AZ91 magnesium alloy in ethylene glycol solution in presence of chloride anions", Materials and Corrosion, 56 (2005) 83-87.

17. K. N. Reichek, K. J. Clark and J. E. Hillis"Controlling the salt water corrosion performance of magnesium AZ91 alloy", SAE Technical Paper series #850417, 1985,1-12.

18. D. Hawke and A. Olsen, "Corrosion properties of new magnesium alloys", Magnesium Properties and Applications for Automobiles, SAE, 1993, SP-962, 79-84.

19. B. E. Carlson and J. W. Jones "The metallurgical aspects of the corrosion behaviour of cast Mg-Al alloys", CIM conference, Quebec, 1993, Light Metals Processing and Applications, 833-847.

20. O. Lunder, K. Nisancioglu and R. Hansen, "Corrosion of die cast magnesium-aluminium alloys", SAE, 1993, SP-93/962, 117-126.

21. F. Lefebvre and G. Nussbaum, "Influence of the microstructure on the corrosion resistance of Mg-Al based alloys", CIM Conference Proceedings, Extraction, Refining and Fabrication of Light Metals, 1991, 20-30.

22. G. Song and D. StJohn, "The effect of zirconium grain refinement on the corrosion behaviour of magnesium-rare earth alloy MEZ", Journal of Light Metals, 2 (2002) 1-16.

23. L. Y. Mei, G. L. Dunlop, and H. Westengen, "Age hardening and precipitation in a cast Mg-rare-earth alloy", Journal of Materials Science, 31 (1996) 387-397.

24. G. Song, A. Atrens and M. Dargusch, "Influence of microstructure on the corrosion of die cast AZ91D", Corrosion Science, 41 (1999) 249-273.

25. S. Mathieu, C. Rapin, J. Steinmetz and P. Steinmetz, "A corrosion study of the main constituent phase of AZ91 magnesium alloys", Corrosion Science, 45 (2003) 2741-2755.

26. G. Song, A. Atrens, X. Wu and B. Zhang, "Corrosion behaviour of AZ21, AZ501 and AZ91 in sodium chloride", Corrosion Science, 40 (1998) 1769-1791.

27. S. Mathieu, C. Rapin, J. Hazan and P. Steinmetz, "Corrosion behaviour of high pressure die-cast and semi-solid cast AZ91D alloys", Corrosion Science, 44 (2002) 2737-2756.

28. M. G. Fontana, "Corrosion Engineering", 3rd Ed., McGraw-Hill International Editions, USA, 1996.

29. G. Song and A. Atrens, "Understanding magnesium corrosion", Advanced Engineering Materials, 5 (2003), 837-858.

30. G. Song and B. Johannesson, S. Hapugoda and D. StJohn, "Galvanic corrosion of magnesium alloy AZ91D in contact with an aluminium alloy, steel and zinc", Corrosion Science, 46 (2004) 955-977.

31. G. L. Makar and J. Kruger, "Corrosion studies of rapidly solidified magnesium alloys", Journal of the Electrochemical Society, 137 (1990) 414-421.

32. F. Czerwinski, "The early stage oxidation and evaporation of Mg-9%Al-1%Zn alloy", Corrosion Science, 46 (2004) 377-386.

33. P. L. Hagans, "Surface modification of magnesium for corrosion protection", in Proc. 41 st World Magnesium Conf., Dayton, OH, 1984. International Magnesium Association, 30-38.

34. S. Akvipat, E. B. Habermann and P. L. Hagnas, "Effects of iron implantation on the aqueous corrosion of magnesium", Materials Science Engineering, 69 (1984), 311-316.

35. Govind, K Nair, M. Mittal, K. Lal, R. Mahanti, and C. Sivaramakrishnan, "Development of rapidly solidified (RS) magnesium-aluminium-zinc alloy", Materials Science and Engineering, A304-306 (2001) 520-523.

36. N. Aung and W. Zhou, "Effect of heat treatment on corrosion and electrochemical behaviour of AZ91D magnesium alloy", Journal of Applied Electrochemistry, 32 (2002) 1397-1401.

37. W. P. Innes, In: Electroplating and Electroless Plating on Magnesium and Magnesium Alloys, Modern Electroplating, Wiley-Interscience, New York, 1974, p602.

38. Y. F. Jiang, C.Q. Zhai, L.F. Liu, Y.P. Zhu, and W.J. Ding, "Zn-Ni alloy coatings pulse-plated on magnesium alloy", Surface and Coatings Technology, 191 (2005) 393-399.

39. A. Sharma, M. Suresh, H. Narayanmurthy and R. P. Sahu "Electroless Nickel Plating on Magnesium Alloy", Metal Finishing, 96 (1998) 10-18.

40. H. Huo, Y. Li and F. Wang, "Corrosion of AZ91D magnesium alloy with a chemical conversion coating and electroless nickel layer", Corrosion Science, 46 (2004) 1467-1477.

41. K. Chong and T. Shih, "Conversion-coating treatment for magnesium alloys by a permanganate-phosphate solution", Materials Chemistry and Physics, 80 (2003) 191-200.

42. D. Hawke and D. Albright,"A Phosphate –Permanganate Conversion Coating for Magnesium", Metal Finishing, 93 (1995) 34-38.

43. H. Umehara, M. Takaya and S. Terauchi, "Chrome-free surface treatments for magnesium alloy", Surface and Coatings Technology, 169-170 (2003) 666-669

44. M. Gonzalez-Nunez, C. A. Nunez-Lopez, P. Skeldon, G. E. Thompson, H. Karimzadeh, P. Lyon and T. E. Wilks"A non-chromate conversion coating for magnesium alloys and magnesium-based metal matrix composites", Corrosion Science, 37 (1995) 1763-1772.

45. A. Rudd, C. Breslin and F. Mansfeld. "The corrosion protection afforded by rare earth conversion coatings applied to magnesium", Corrosion. Science, 42 (2000) 275-288.

46. J. E. Hillis, "Surface engineering of magnesium alloys", in: ASM Handbook, Surface Engineering, Vol. 5, ASM International, 1994.

47. H. A. Evangelides, "A new finish for magnesium alloys", Metal Finishing, 55 (1951) 56-60.

48. Anon, "The HAE process to date", Light Metal Age, 15 (1957) 10-14.

49. H. Umehara, S. Terauchi and M. Takaya, "Structure and corrosion behaviour of conversion coatings on magnesium alloys", Materials Science Forum, 350 (2000) 273-282.

50. Y. Mizutani, S. J. Kim, R. Ichino and M. Okido, "Anodizing of Mg alloys in alkaline solutions", Surface and Coatings Technology, 169-170 (2003) 143−146.

51. R. Bunshah, "Handbook of Hard Coatings", William Andrew Publisher, New York, USA, 2001.

52. F. Fracassi, R. Agostino, F. Palumbo, E. Angelini, S. Grassini and F. Rosalbino"Application of plasma deposited organodilicon thin films for the corrosion protection of metals", Surface and Coatings Technology, 174−175 (2003) 107−111.

53. N. Yamauchi, K. Demizu, N. Ueda, N.K. Cuong, T. Sone and Y. Hirose, "Friction and wear of DLC films on magnesium alloy", Surface and Coatings Technology, 193 (2005) 277−282.

54. F. Hollstein, R. Eiedemann and J. Scholz, "Characteristics of PVD-coatings on AZ31hp magnesium alloys", Surface and Coatings Technology, 162 (2003) 261-268

55. H. Hoche, H. Scheerer, D. Probst, E. Broszeit and C. Berger, "Development of a plasma surface treatment for magnesium alloys to ensure sufficient wear and corrosion resistance", Surface and Coatings Technology, 174−175 (2003) 1018−1023.

56. A. Phani, F. Gammel, T. Hack, and H. Haefke, "Enhanced corrosion resistance by sol-gel-based ZrO2-CeO2 coatings on magnesium alloys", Materials and Corrosion, 56 (2005) 77-82.

57. A. L. Yerokhin, X. Nie, A. Leyland, A. Matthews and S. J. Dowey"Plasma electrolysis for surface engineering", Surface and Coatings Technology, 122 (1999) 73-93.

58. A. L. Yerokhin, V.V. Lyubimov and R.V. Ashitkov, "Phase formation in ceramic coatings during plasma electrolytic oxidation of aluminium alloys", Ceramics International, 24 (1998) 1−6.

59. Y. Wang, J. Wang and J. Zhang, "Characteristics of anodic coatings oxidized to different voltage on AZ91D Mg alloy by micro-arc oxidization technique", Materials and Corrosion, 56 (2005) 88−92.

60. O. Khaselev, D. Weiss and J. Yahalom, "Structure and composition of anodic films formed on binary Mg-Al alloys in KOH-aluminate solutions under continuous sparking", Corrosion Science, 43 (2001) 1295-1307.

61. Y. Zhang, C. Yan, F. Wang, H. Lou and C. Cao, "Study on the environmentally friendly anodizing of AZ91D magnesium alloy", Surface and Coatings Technology, 161 (2002) 36-43

62. H. Hsiao and W. Tsai, "Characterization of anodic films formed on AZ91D magnesium alloy", Surface and Coatings Technology, 190 (2005) 299-308

63. W. B. Xue, Z. W. Deng, T. H. Zhang, R. Y. Chen and Y. L. Li, "Microarc oxidation mechanism of a cast magnesium alloy", Rare Metal Materials Engineering, 28 (1999) 353-356.

64. A. L. Yerokhin, A. Ahatrov, V. Samsonov, P. Shashkov, A. Leyland and A. Matthews. "Fatigue properties of Keronite® coatings on a magnesium alloy", Surface and Coatings Technology, 182 (2004) 78−84.

65. X. W. Guo, W.J. Ding, C. Lu and C.Q. Zhai, "Influence of ultrasonic power on the structure and composition of anodizing coatings formed on Mg alloys", Surface and Coatings Technology, 183 (2004) 359−368.

66. H. Tamai, K. Igaki, E. Kyo, Initial and 6-Month Results of Biodegradable Poly-L-Lactic Acid Coronary Stents in Humans, Circulation, 102, 2000, 399–404.

67. P. Zhang, X. Nie, H. Hu, Liu, Y., TEM analysis and Tribological Properties of Plasma Electrolytic Oxidation (PEO) Coatings on a Magnesium Engine AJ62 Alloy, Journal of Surface and Coatings Technology, Vol 205, 2010, 1508-1514.

68. L. Han, X. Nie, Q. Zhang, H. Hu, Influence of Electrolytic Plasma Oxidation Coating on Tensile Behavior of Die Cast AM50 Alloy Subjected to Salt Corrosion, International Journal of Modern Physics B, Vol 23 (6&7), 2009, 960-965.

69. Y. Ma, H. Hu, D. Northwood, X. Nie, Optimization of the Electrolytic Plasma Oxidation Processes for Corrosion Protection of Magnesium Alloy AM50 Using the Taguchi Method, Journal of Materials Processing Technology, Vol 182, 2007, 58–64.

70. Y. Ma, X. Nie, D.O. Northwood, H. Hu, Systematic Study of the Electrolytic Plasma Oxidation Process on Mg Alloy for Corrosion Protection, Solid Thin Films, Vol 494, 2006, 296-301.

71. Y. Ma, X. Nie, D. O. Northwood, H. Hu, Corrosion and Erosion Properties of Oxide Coatings on Magnesium, Thin Solid Films, Vol 469-470, 2004, 472-477.

72. W. Zhang, B. Tian, K. Du, H. X. Zhang, F. H. Wang, Preparation and Corrosion Performance of PEO Coating With Low Porosity on Magnesium Alloy AZ91D In Acidic KF System, International Journal of Electrochemical Science, 6, 2011, 5228-5248.

73. R.O. Hussein, P. Zhang, D.O. Northwood, and X. Nie, Improving the Corrosion Resistance of Magnesium Alloy AJ62 by a Plasma Electrolytic Oxidation (PEO) Coating Process, Corrosion & Materials, Vol. 36, No. 3, 2011, 38-49.

74. B. Denkena, A. Lucas, Biocompatible Magnesium Alloys as Absorbable Implant Materials –Adjusted Surface and Subsurface Properties by Machining Processes, Annals of the CIRP, 56 (1), 2007, 113-116.

75. B. Denkena, F. Witte, C. Podolsky, A. Lucas, Degradable Implants Made of Magnesium Alloys, Proc. of the 5th euspen International Conference – Montpellier – France, May 2005.

76. G. Song, S. Song, A Possible Biodegradable Magnesium Implant Material, Advanced Engineering Matierals, 9, No. 4, 2007, 298-302.

77. J. Li, P. Hanb, W. Ji, Y. Song, S. Zhang, Y. Chen, C. Zhao, F. Zhang, X. Zhang, Y. Jiang, The In Vitro Indirect Cytotoxicity Test and In Vivo Interface Bioactivity Evaluation of Biodegradable FHA Coated Mg-Zn Alloys, Materials Science and Engineering B, 176, 2011, 1785– 1788

78. J. Hu and X. Nie, "Plasma Electrolytic Oxidation Treatment of Pure Magnesium for Potential Biological Application", Proceedings of Biomaterials, Smart Materials, and Structures, The 8th Pacific Rim International Congress on Advanced Materials, TMS, August 4-9, 2013, Waikoloa, Hawaii, USA, Page 1655-1662.

Chapter 8

Corrosion Protection of Magnesium Alloys in Industrial Solutions

Amany Mohamed Fekry[1]

[1]Cairo University (Faculty of Science, Chemistry Department), Egypt

INTRODUCTION

The main problem in our life is the corrosion of many types of alloys either industrialy or biologicaly. This work reviews the corrosion protection of magnesium based alloys in industrial solutions. Corrosion behavior had been studied using electrochemical impedance spectroscopy (EIS), Potentiondynamic polarization and scanning electron microscope (SEM) techniques. Magnesium is the lightest of all metals in practical use with density of 1.74 g cm^{-3}. Pure magnesium metal has useful properties such as shielding against electromagnetic waves, vibration damping, dent resistance and machinability, in addition to its recyclability as it has a lower specific heat and a lower melting point than other metals. On the other hand, magnesium has shortcomings such as insufficient strength, elongation and heat resistance as well as being subject to corrosion. It is necessary to deal with its shortcomings and improve its performance through alloying with various elements. Alloying magnesium improves its strength, heat resistance and creep resistance [1].

Magnesium alloys are the most versatile and attractive metallic materials. They are used for a broad range in commercial, industrial and aerospace applications due to their many advantages, such as light density, good mechanical properties and excellent castability [2-4]. The most common magnesium alloys are those containing aluminum, Zinc, manganese, zirconium, thorium and rare earth metals. The latter alloys are used when resistance to creep and high temperature strength are required. One of the major problems that limit magnesium alloys application is their high susceptibility to corrosion in different media [5] which depends widely on film formation and surface electrolyte interaction. A serious limitation for the wide-spread use of several magnesium alloys is their susceptibility to general and localized (pitting) corrosion. The AZ-based Mg system has been the basis of the most widely used magnesium alloys [6]. One of the most successful magnesium-aluminium alloy is AZ91D, which has a two-phase microstructure typically consisting of α-Mg matrix with the β-phase (the intermetallic $Mg_{17}Al_{12}$) distributed along the α grain boundaries. The α-phase consists of α-Mg-Al-Zn solid solution with the same structure as pure magnesium [5]. Also AZ31E alloy have excellent mechanical properties. Extruded Mg alloys as AZ31E is getting more and more widely used because of their considerably high plasticity in comparison with the die-cast Mg alloys [7].

Corrosion is a major problem in the cooling system of an engine block. Currently, the main composition of a conventional coolant is 30–70 vol% ethylene glycol [7]. This can be used for studying the corrosion behavior of AZ91D alloy. Most existing commercial coolants fail to provide adequate corrosion protection to magnesium alloys [8]. Some companies have also realized the difficulty of using the conventional coolants for magnesium alloys, and are developing coolants with new inhibitors [9]. Song et al. [7] observed that the corrosion rate of magnesium in aqueous ethylene glycol depends on the concentration of the solution. A diluted ethylene glycol solution is more corrosive than a concentrated one at room temperature. Ethylene glycol solution contaminated by individual contaminants NaCl, NaHCO$_3$ or Na$_2$SO$_4$ is more corrosive to magnesium. NaCl is the most detrimental contaminant. In this work, a study for the corrosion behavior of AZ91D alloy in this coolant has been done, due to this issue is important industrially. The aim is to characterize the corrosion properties of AZ91D alloy in aqueous ethylene glycol-water solutions of different percentages. The influence of adding chloride or flouride ion in ethylene glycol solution on the corrosion behavior was studied. Also the effect of paractamol ((N-acetyl-para-aminophenol=APAP) as inhibitor in ethylene glycol for AZ91D alloy is investigated [6].

Electrochemical characterization and corrosion behavior of AZ31E alloy was done in aqueous Oxalic acid as industrial solution [10]. Oxalic acid is a relatively strong organic acid used as purifying agent in pharmaceutical industry. Oxalic acid's main applications include cleaning or bleaching, especially for the removal of rust [11]. The work aims to attain more information concerning the corrosion behavior of AZ31E alloy in oxalic acid solution containing Cl$^-$, F$^-$or PO$_4^{3-}$ anions under various environmental conditions. The corrosion rate was found to increase with increasing oxalic acid concentration. The effect of adding Cl$^-$, F$^-$or PO$_4^{3-}$ ions on the electrochemical behavior of AZ31E electrode was studied in 0.01 M oxalic acid solution at 298 K. It was found that the corrosion rate increases with increasing Cl$^-$or F$^-$ion concentration, however, it decreases with increasing PO$_4^{3-}$ ion concentration [12].

THE MAIN PROBLEM

One of the major problems that limit magnesium alloys application is their high susceptibility to corrosion in different media. This makes studying the corrosion and corrosion control of Mg alloys an interesting point of research which can enable extending the potential use of these important materials in a broad range of many technical and innovative applications.

THE AIM

The aim is to study the corrosion resistance of Mg alloys which depends essentially on two main factors: (i) alloy microstructure and (ii) properties of the developed surface film.

Generally, it is aimed to find the best magnesium alloy with low cost and low corrosion rate and to find a possible way to improve corrosion resistance of either AZ91D or AZ31E alloy in different industrial solutions.

EXPERIMENTAL

Samples of die cast magnesium aluminum alloy (AZ91D or AZ31E) in the form of plates (200 x 100 x 5 mm) were donated from Department of mining, Metallurgy and Materials Engineering, Laval University, Canada. The chemical composition (wt%) of the two alloys are as follows: 9.0 Al, 0.67 Zn, 0.33 Mn, 0.03 Cu, 0.01 Si, 0.005 Fe, 0.002 Ni, 0.0008 Be and balance Mg for AZ91D alloy; and 2.8 Al, 0.96 Zn, 0.28 Mn, 0.0017 Cu, 0.0111 Fe, 0.0007 Ni, 0.0001 Be and balance Mg for AZ31E. They were used for preparing the working electrodes. The sample was divided into small coupons. Each coupon was welded to an electrical wire and fixed with Araldite epoxy resin in a glass tube leaving cross-sectional area of the specimen 0.2 cm² for AZ91D alloy and 0.196 cm² for both AZ31E alloy.

The solutions used were prepared using Analar grade reagents for each work are (ethylene glycol, sodium fluoride, sodium chloride and paracetamol [6]) and (oxalic acid, sodium fluoride, sodium chloride and sodium phosphate [12]). All solutions were prepared using triply distilled water.

The surface of the test electrode was mechanically polished by emery papers with 400 up to 1000 grit to ensure the same surface roughness, degreasing in acetone, rinsing with ethanol and drying in air.

The cell used was a typical three-electrode one fitted with a large platinum sheet of size 15×20×2mm as a counter electrode (CE), saturated calomel (SCE) as a reference electrode (RE) and the alloy as the working electrode (WE).

The impedance diagrams were recorded at the free immersion potential (OCP) by applying a 10 mV sinusoidal potential through a frequency domain from 100 kHz down to 100 mHz. The EIS was recorded after reading a steady state open-circuit potential. The polarization scans were carried out at a rate of 1 mV/s over the potential range from-2.5 to 0 mV vs. saturated calomel electrode (SCE). Prior to the potential sweep, the electrode was left under open-circuit in the respective solution until a steady free corrosion potential was recorded. Corrosion current, i_{corr}, which is equivalent to the corrosion rate is given by the intersection of the Tafel lines extrapolation. In the weight loss measurements, the treated samples were weighed before and after the immersion in Hank's solution in absence and in presence of different concentrations of glucosamine sulphate. The instrument used is the electrochemical workstation IM6e Zahner-elektrik, GmbH, (Kronach, Germany).The electrochemical and weight loss experiments were always carried inside an air thermostat which was kept at 25°C, unless otherwise stated. The SEM micrographs were collected using a JEOL JXA-840A electron probe microanalyzer.

RESULTS AND DISCUSSION

Electrochemical Impedance Measurements

AZ91D Alloy in Ethylene Glycol Solution [6]

The EIS scans of AZ91D alloy as a function of concentration for ethylene glycol were recorded (Figure 1) after leaving the working electrode for 2 h in the test solution until reaching a steady state potential value (E_{st}). As shown in Figure 1, an increase in ethylene glycol concentration leads to a decrease in the |Z| value, indicating that pure ethylene glycol is almost inert to magnesium alloy and the corrosion of magnesium alloy in ethylene glycol solution is closely related to the water content of the solution [7].

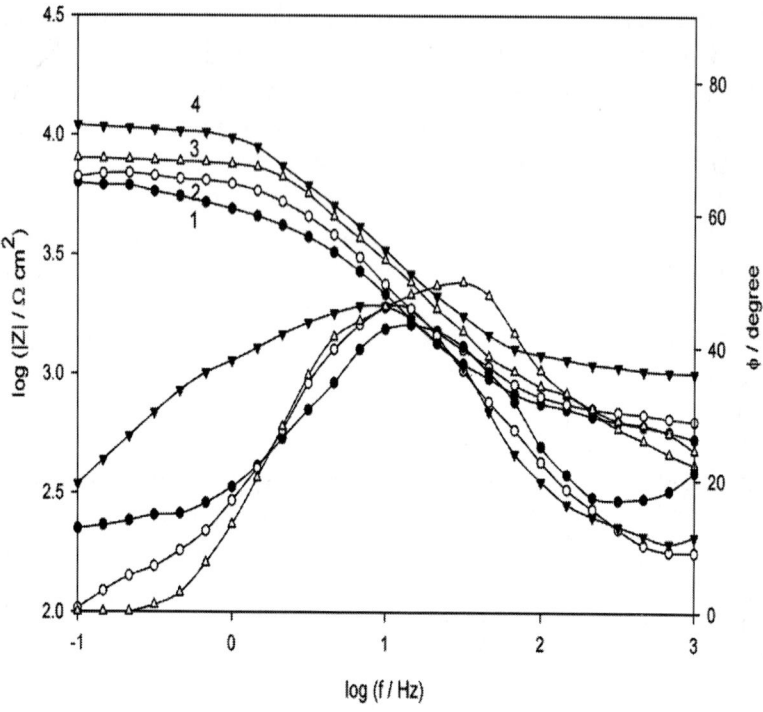

Figure 1: EIS data of AZ91D alloy exposed after 2 h immersion in various concentrations of ethylene glycol solution: (1) 30%, (2) 50%, (3) 70% and (4) 90%.

By studying the effect of adding either fluoride or chloride to the highest corrosive concentration (30% ethylene glycol-70% water), it was found that the impedance value decreases (Figure 2a) with increasing chloride ion concentration due to its aggressiveness [10]. However, for fluoride containing ethylene glycol solution impedance value increases with increasing F⁻ion concentrations shown in Figure 2b.

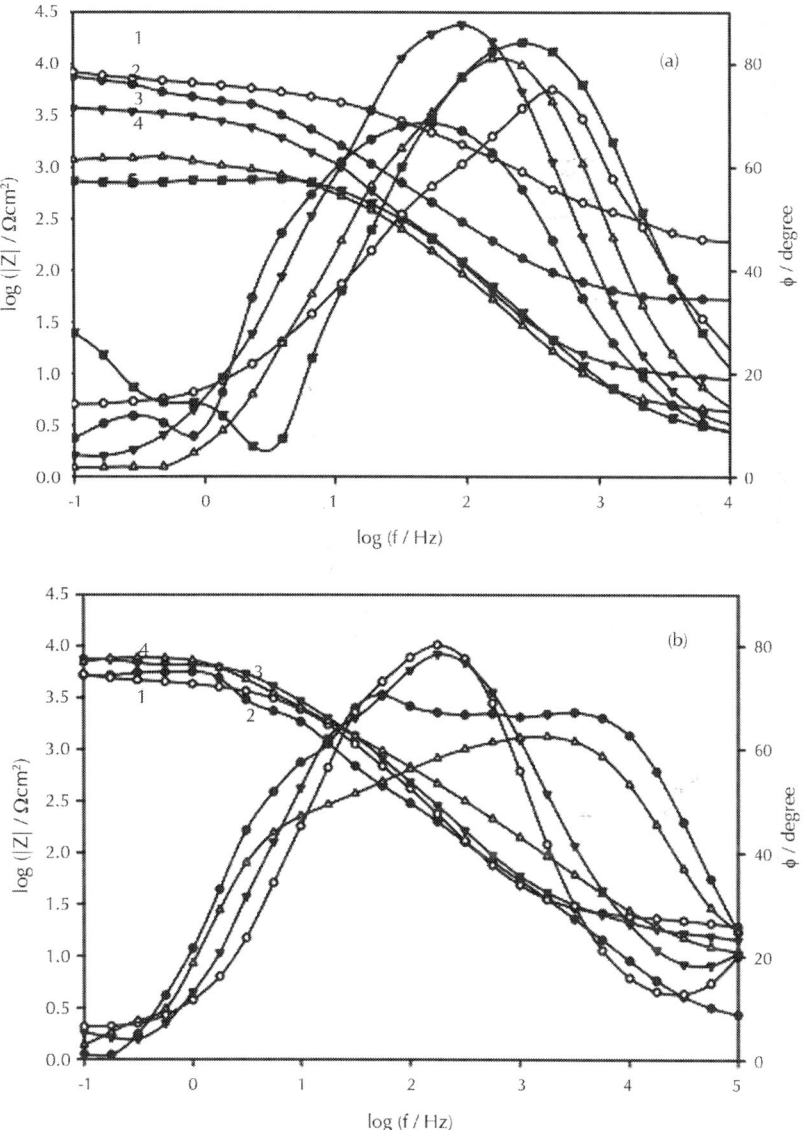

Figure 2: EIS data of AZ91D alloy exposed after 2 h immersion in 30% ethylene glycol solution with (a) chloride and (b) fluoride ions of various concentrations: (1) 0.01 M, (2) 0.05 M, (3) 0.1 M, (4) 0.3 M and (5) 0.6 M.

On adding paracetamol as inhibitor in the concentration range (0.01-1.0 mM) for corrosion of the blank (30% ethylene glycol-70%

water), it was found that both |Z| value and phase angle maximum (φ) increase suddenly up to 0.05 mM then decrease regularly up to the highest concentration of inhibitor as shown in Figure 3(a,b) as Bode and Nyquist formats, respectively.

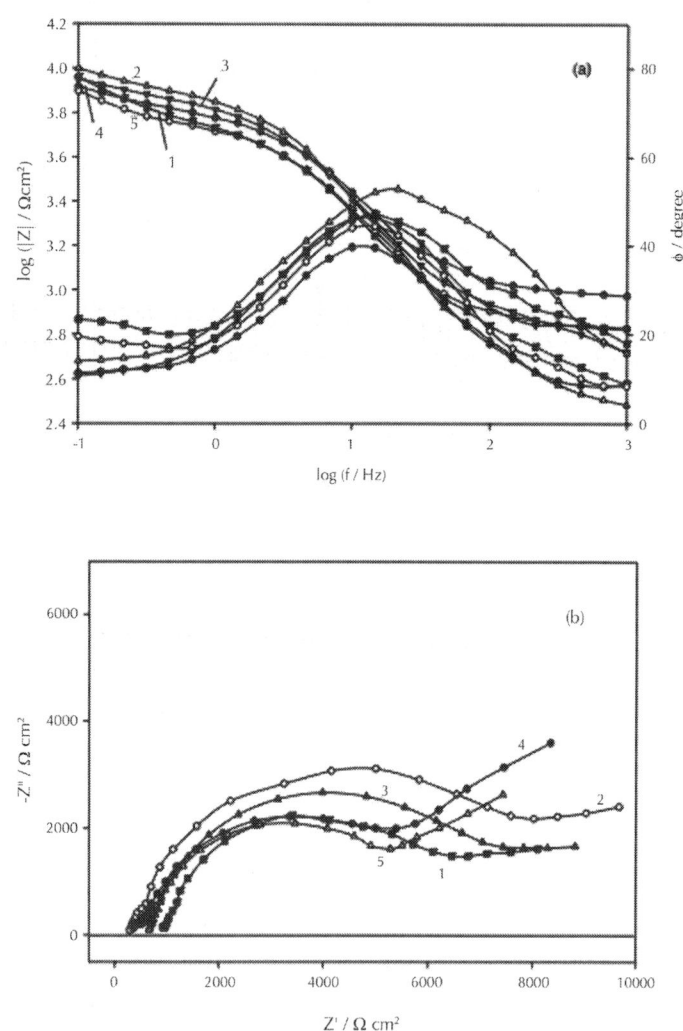

Figure 3: EIS data (a) Bode plots and (b) Nyquist plots of AZ91D alloy exposed after 2 h immersion in 30% ethylene glycol solution with paracetamol of various concentrations: (1) 0.01 mM, (2) 0.05 mM, (3) 0.1 mM, (4) 0.5 mM and (5) 1.0 mM.

The results in general reveal two clear trends concerning the number of peaks observed in the patterns of the phase shift. The first one is for the behavior of AZ91D alloy in chloride, flouride and paracetamol containing ethylene glycol, where the Bode plots display only one maximum phase lag at all tested concentrations, however, for paracetamol, the phase angle maximum is nearly 45°, corresponding to a diffusion control in the passive layer. The second trend is for the alloy behavior in ethylene glycol medium with different concentrations where another peak of phase lag appears at the low frequency region and also the phase angle maximum is nearly 45° due to diffusion phenomenon. The impedance data were thus simulated to the appropriate equivalent circuit for the cases with one time constant (Figure 4a,b) and the others exhibiting two time constants (Figure 4c), respectively. This simulation gave a reasonable fit. The estimated data for ethylene glycol is given in Table 1, for chloride and fluoride ions in Table 2 and for paracetamol ethylene glycol containing solution in Table 3.

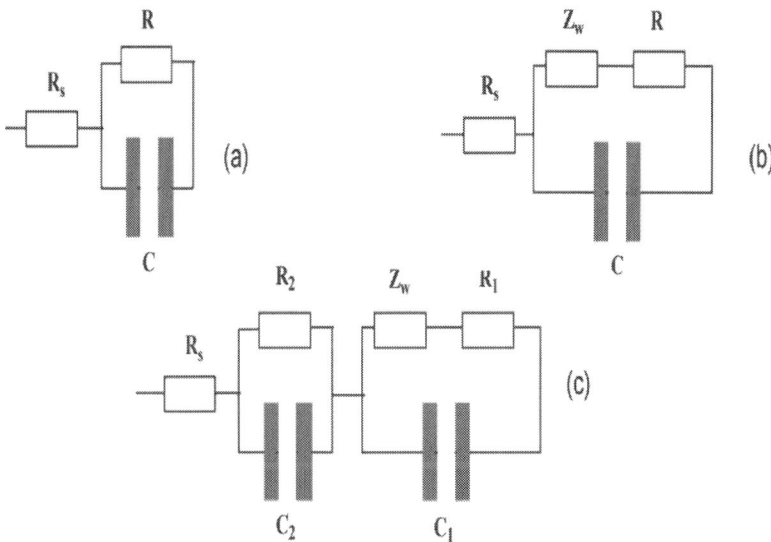

Figure 4: The equivalent circuit model representing (a,b) one and (c) two time constants.

Table 1: Equivalent circuit and corrosion parameters for AZ91D alloy in various concentrations of ethylene glycol solution after 2 h immersion

Cethylene glycol (vol.%)	R^S (Ω cm²)	R^1 (kΩ cm²)	C^1 (nF cm⁻²)	R^2 (kΩ cm²)	C^2 (μF cm⁻²)	W (kΩ cm²s-1/2)	R^T (kΩ cm²)	$1/C^T$ (μF⁻¹cm⁻²)	i_{cor}^r (μA cm⁻²)	E_{cor}^r V
30	11.4	0.5	88.9	4.4	4.5	1.64	5.0	11.5	0.50	-1.45
50	21.9	0.7	77.9	6.2	4.3	1.52	6.9	13.1	0.41	-1.34
70	26.8	1.2	71.0	10.5	4.2	1.47	11.7	14.3	0.25	-1.26
90	30.0	1.4	66.4	15.4	4.0	1.33	16.8	15.3	0.1	-1.18

Table 2: Equivalent circuit and corrosion parameters for AZ91D alloy after 2 h immersion in 30% ethylene glycol solution with different concentrations of Cl⁻ or F⁻ ions

Salt	C_{anio}^n (M)	R (kΩ cm²)	C (µF cm⁻²)	R^s (Ω cm²)	i_{cor}^r (µA cm⁻²)	E_{cor}^r (V)
NaCl	0.01	7.7	2.67	154.8	0.47	-1.36
	0.05	5.8	4.36	50.2	0.49	-1.43
	0.10	3.1	7.44	9.0	0.64	-1.52
	0.30	1.1	8.50	4.2	3.40	-1.53
	0.60	0.6	9.58	2.5	30.4	-1.63
NaF	0.01	4.1	3.54	22.1	0.19	-1.41
	0.05	5.5	3.46	21.8	0.11	-1.43
	0.10	7.0	2.46	15.8	0.09	-1.45
	0.30	8.4	1.23	7.3	0.03	-1.51

Table 3: Equivalent circuit and corrosion parameters for AZ91D alloy after 2 h immersion in 30% ethylene glycol solution with different concentrations of paracetamol

Cparacetamol	R^S	R	C	W	i_{cor}^r	E_{cor}^r	IE
mM	(kΩ cm^2)	(kΩ cm^2)	(μF cm^{-2})	kΩ cm$^{2s-1/2}$	(μA cm^{-2})	V	%
0.01	0.33	6.1	2.7	2.90	0.100	-1.37	80.0
0.05	0.37	10.0	2.2	1.84	0.020	-1.26	96.0
0.1	0.47	8.5	2.4	1.39	0.040	-1.41	92.1
0.5	0.64	5.6	2.8	2.52	0.045	-1.44	91.0
1.0	0.92	5.1	3.6	1.46	0.051	-1.46	89.8

Generally, the impedance response is well simulated by the classic parallel resistor capacitor (RC) combination in series with the solution resistance (R_s). In this model [10] a charge transfer resistance (R) is in parallel with the double layer capacitance (C), as shown in Figure 4a. Figure 4b is one time constant model containing Warburg impedance (Z_w) in series to R [13], which is related to ion diffusion through passive film and indicates that the corrosion mechanism is controlled not only by a charge-transfer process but also by a diffusion process. The appropriate equivalent model for the impedance diagrams with two time constants, consists of two series circuits, $R_1 Z_w C_1$ and $R_2 C_2$ parallel combination and both are in series with R_s. C_1 is the capacitance of the outer layer, C_2 pertains to the inner layer, while R_1 and R_2 are the respective resistances of the outer and inner layers constituting the surface film, respectively [14]. A linear region at the lower frequencies in the Nyquist plot in Figure 3b would be related to diffusion phenomena [15, 16] thereby an equivalent circuit with Warburg component Z_w is more appropriate. Analysis of the experimental spectra were made by best fitting to the corresponding equivalent circuit using Thales software provided with the workstation where the dispersion formula suitable to each model was used [14]. In this complex formula an empirical exponent (a), varying between 0 and 1, is introduced to account for the deviation from the ideal capacitive behavior due to surface inhomogeneties, roughness factors and adsorption effects [10]. An ideal capacitor corresponds to $a=1$ while $a=0.5$ becomes the CPE in a Warburg component [17]. In all cases, good conformity between theoretical and experimental results was obtained for the whole frequency range with an average error of 5%.

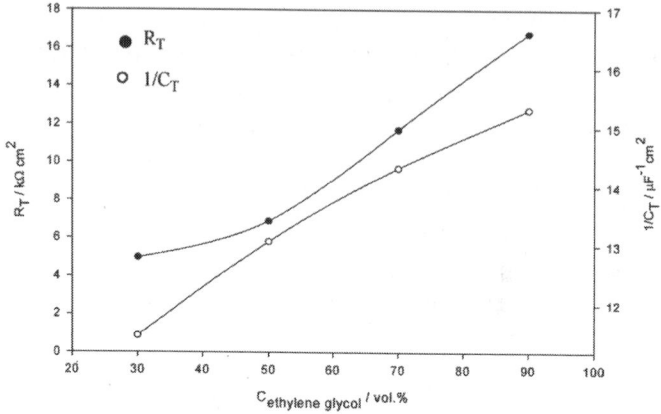

Figure 5: The total resistance (R_T) and relative thickness ($1/C_T$) for AZ91D alloy at various concentrations of ethylene glycol solution, measured after 2 h immersion.

The effect of concentration for ethylene glycol or additive ions or inhibitor on the relative thickness ($1/C_T$) [18] of AZ91D. Figure 5 reveals features generally concurrent to the behavior of the film resistance. It shows that the resistance (R_T) and the relative thickness ($1/C_T$) of the surface film on AZ91D sample increase with increasing the concentration of ethylene glycol. Thus, 30% ethylene glycol (blank) is an aggressive solution as shown in SEM image in Fig. 6a, where corrosion products appear on the surface. Pure ethylene glycol has very poor electrical conductivity and is almost an insulator [7]. However, dilution by water facilitates the hydrolysis of the hydroxyl groups in ethylene glycol increasing its electrical conductivity. Ethylene glycol molecule is larger than water, so the adsorption of the former at the surface of AZ91D alloy can result in a lower capacitance value [7]. When the concentration of ethylene glycol increases, more ethylene glycol will be adsorbed on the surface, leading to a lower C_T. This explains the decreasing corrosion rate of AZ91D alloy with increasing concentration of ethylene glycol and also the decrease in the Warburg impedance diffusion.

(a)   (b)

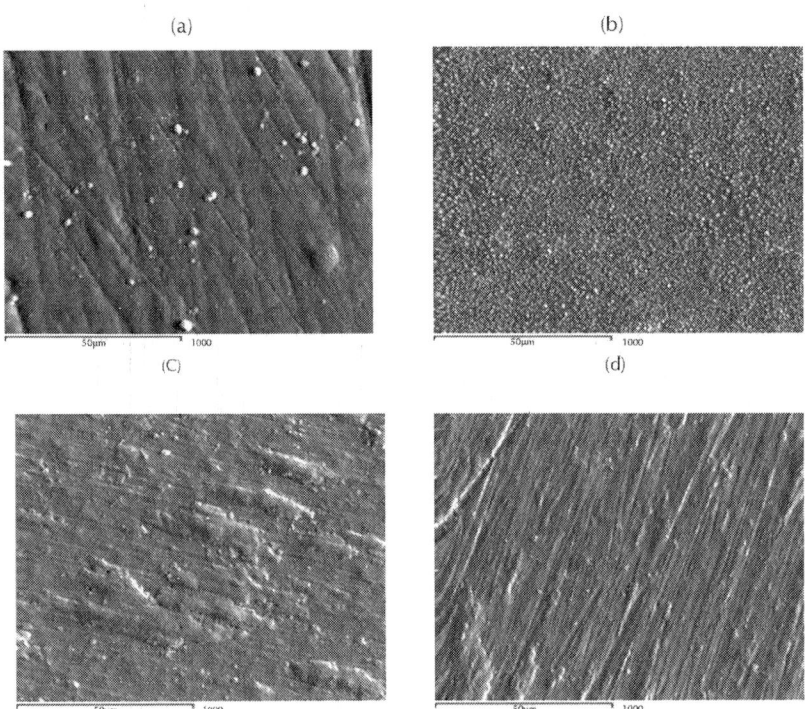

(C)   (d)

Figure 6: (a-d). SEM micrograph of the (a) blank (30% ethylene glycol solution), (b) 0.3 M F-(c) 0.05 mM paracetamol and (d) 1.0 mM paracetamol, ethylene glycol containing solution.

As given in Table 2, in ethylene glycol solution contaminated with chloride ions, the resistance decreases sharply at first then reaches a quasi-state value with increasing concentration of contaminant [6]. The addition of chloride does not significantly increase the capacitance value until the amount of the added Cl-ions is above a certain level (> 0.05 M). At concentrations > 0.05 M, they are more corrosive than the blank, indicating film dissolution, which can be attributed to the more aggressive nature of the chloride anion. The increase in capacitance should be due to the replacement of ethylene glycol on the alloy surface by the chloride ions as contaminant.

Fluoride is also an important substance that could exist in normal water and can easily be introduced into vehicle coolant systems [19]. As given in Table 2, in fluoride medium at concentrations > 0.05 M, R and 1/C of the surface film increase steeply than the blank due to the formation of less soluble and more stable magnesium fluoride (MgF_2) film [17]. This is confirmed by SEM micrograph of 0.3 M fluoride in ethylene glycol concentration (Figure 6b), the grain particles of the salt film grow laterally during the prolonged exposure (2 h) covering nearly the whole surface of the alloy indicating more stability as compared to the blank shown in Figure 6a. In fact, F-has recently been used as an inhibitor in coolants for magnesium alloys [7]. However, the inhibition mechanism has not been systematically studied. Gulbrandsen et. al. [20] reported that crystalline $KMgF_3$ was identified on magnesium at higher F-concentration in the more alkaline solutions. At 0.01 M fluoride concentration ethylene glycol may help in the formation of this compound, so that the resistance decreases than the blank. Table 2 shows that the addition of F-into ethylene glycol strikingly enhanced R but decreased R_s and C. The significantly reduced C and dramatically improved R suggest that [7] a three-dimensional film was formed on the magnesium surface which is much thicker than the adsorbed film and thus can effectively separate the magnesium alloy from the solution, making the corrosion reaction at the interface very slow. As to the solution resistance R_s, its decrease after the addition of F-can be simply attributed to the increased total concentration of ions by adding F-into the solution.

Finally, the effect of adding paracetamol as inhibitor was studied; it was found that all concentrations give good inhibition as compared to the blank which may be due to the adsorption of the inhibitor through the adsorption. The rate of adsorption is usually rapid and hence, the reactive metal surface is shielded from the aggressive environment [21]. However, it was found that there is a critical concentration for the inhibitor at 0.05 mM which has the highest resistance as shown in Table 3, and the resistance decreases with increasing the inhibitor concentration > 0.05 mM. This behavior is confirmed by SEM micrographs shown in Figure 6(c,d), where Figure 6c is for 0.05 mM and Figure 6b for 1.0 mM paracetamol containing ethylene glycol solution. Figure 6c shows a denser and smoother film adsorbed on the alloy surface than that formed on 1.0 mM concentration Figure 6d, Also the two are much better than the blank shown in Figure 6a.

Potentiodynamic Polarization Measurements

AZ91D Alloy in Ethylene Glycol Solution [6]

The anodic and cathodic (E-log i) plots of AZ91D alloy in the ethylene glycol solution of different concentrations were also studied using potentiodynamic polarization measurements at a scan rate of 1.0 mV s^{-1}. The curves were swept from-2.5 V to-1.0 V vs. SCE. Prior to the potential scan the electrode was left for 2 h until a steady free corrosion potential (E$_{st}$) value was recorded. The electrochemical parameters shown in Tables 1 and 2 were obtained by analyzing the I/E data as described elsewhere [10]. The corrosion potential (E$_{corr}$) and current density (i$_{corr}$) were calculated by Tafel extrapolation method for the cathodic branches of the polarization curves. Furthermore, to illustrate the relative stability of the surface film on AZ91D alloy in the investigated solutions, i$_{corr}$ values are found to decrease and E$_{corr}$ values shifts positively with increasing ethylene glycol percentage. Since increasing water percentage in ethylene glycol is responsible for the corrosivity of the solution to the alloy. However, in chloride containing solution i$_{corr}$ values increases and E$_{corr}$ values shifts to more negative values with increasing Cl$^-$concentration. This behavior reflects the harmful influence of Cl$^-$ions on the corrosion performance of AZ91D in aqueous liquids [7]. In Flouride containing solutions, a strange behavior occurs, where i$_{corr}$ decreases and E$_{corr}$ values tend to more negative values with increasing F$^-$concentration. Particularly, the role of the β-phase in corrosion is extensively addressed for AZ91D, and it is generally accepted that the β-phase is a corrosion barrier and its presence in an AZ91D alloy is beneficial to the corrosion resistance of the alloy. The reason is fluoride refined AZ91D magnesium alloy by blocking the growth of primary fir-tree crystals in the crystal boundary [22]. Thus, the dimension of β phase is decreased. In the cathodic reaction process, the overpotential of the hydrogen generation increased due to the dispersion of β phase, which resulted in the corrosion potential of the AZ91D.

Fig. 7 shows polarization scans for paracetamol, from which i$_{corr}$ values is calculated and drawn against inhibitor concentration (Fig. 8). As can be seen, i$_{corr}$ value is the lowest at 0.05 mM paracetamol concentration, which is a critical concentration and shows the highest inhibition efficiency (IE) of 96% which calculated from the following equation:

$$IE\% = 1 - \frac{i_{inh}}{i_{corr}} \times 100 \tag{1}$$

Generally, impedance and polarization measurements confirm each other and all are confirmed by SEM images.

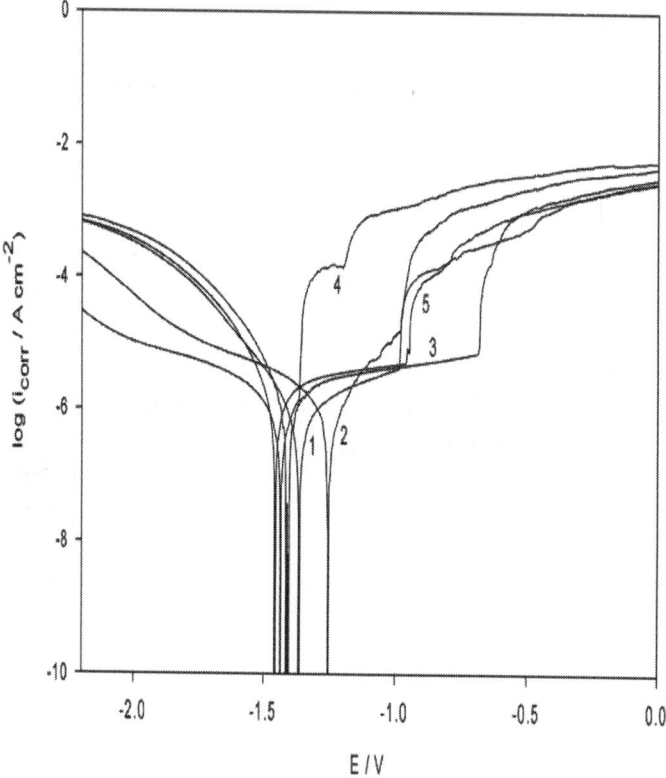

Figure 7: Cathodic and anodic scans of AZ91D alloy exposed after 2 h immersion in 30% ethylene glycol solution with paracetamol of various concentrations: (1) 0.01 mM, (2) 0.05 mM, (3) 0.1 mM, (4) 0.5 mM and (5) 1.0 mM.

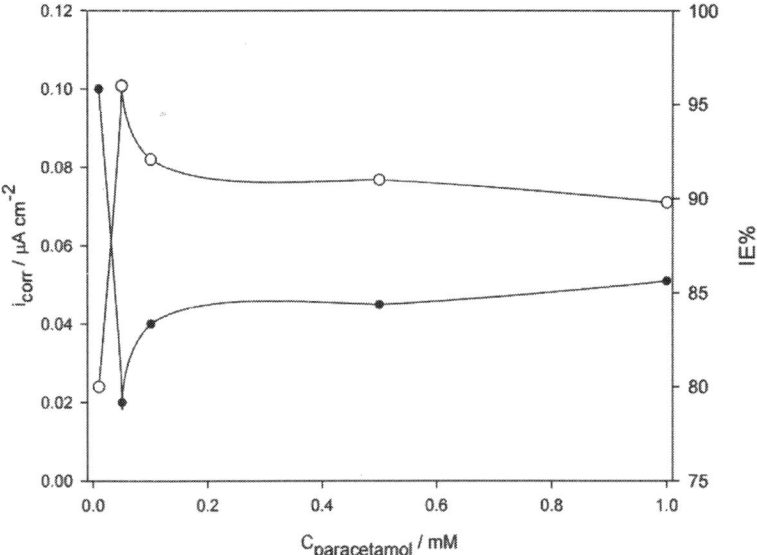

Figure 8: Variation of i_{corr} and IE% for AZ91D alloy exposed after 2 h immersion in 30% ethylene glycol solution with various concentrations of paracetamol.

AZ31E Alloy in Oxalic Acid Solution [12]

The impedance measurements recorded after 2 hours of immersion for AZ31E electrode in oxalic acid solution with different concentrations are presented in Figure 9. Bode plots show an intermediate frequency phase peak shifts to lower frequency and higher phase angle maximum with decreasing oxalic acid concentration. Also, impedance values increase with decreasing the concentration of oxalic acid. The appropriate equivalent model used (Figure 4c) consists of two circuits in series from $R_1 C_1 Z_W$ and $R_2 C_2$ parallel combination and both are in series with R_s as discussed above. In all cases, good conformity between theoretical and experimental results was obtained with an average error of 4%. The evaluated experimental values are given in Table 4.

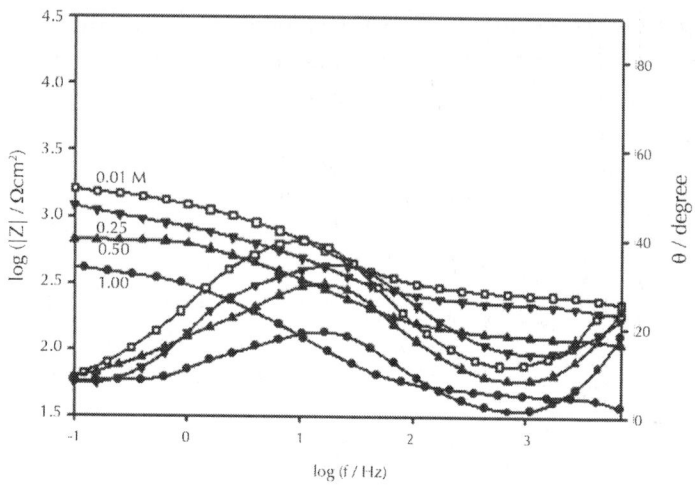

Figure 9: Bode plots of AZ31E electrode in naturally aerated oxalic acid solution of different concentrations, at 298 K.

Table 4: Impedance parameters of AZ31E electrode in naturally aerated oxalic acid of different concentrations, at 298 K

Na$_2$C$_2$O$_4$	R^1	C^1	1	R^2	W	C^2	2	R^S
M	kΩ c^{m}2	μF c^{m-2}		Ω c^{m2}	kDW	μF c^{m-2}		Ω c^{m2}
0.01	2.1	4.7	0.94	58.3	10.2	19.2	0.58	243
0.25	1.3	5.4	0.93	35.6	9.7	20.1	0.56	200
0.50	0.8	6.6	0.91	22.3	7.5	21.3	0.55	100
1.00	0.7	8.3	0.90	14.4	5.2	22.0	0.52	44

It was found that, when AZ31E electrode was immersed in oxalic acid solution, two competitive processes occur. The first one is oxide formation which yields a compact magnesium oxide film with good corrosion resistance. The second one is the formation of magnesium oxalate complexes, which yields a thick porous film as in case of Aluminum alloys [23] with expected low corrosion resistance, where oxalate ions are bidentate ligands capable of forming strong surface complexes. With increasing of oxalic acid concentration the alternation of the compact oxide film by porous one will increase leading to an increase in the corrosion rate. This also is due to increasing of the acidity of the medium.

In Table 4, R_1 represents the resistance of the passive film which decreases with increasing of oxalic acid concentration due to alternation of compact film by porous one. Consequently, the decrease in the relative thickness of the passive film ($1/C_1$) supports this concept. As the most stable formula for magnesium oxalate is dehydrated one [24], so R_2 can represent the resistance of the hydrated layer and the decreasing of relative thickness ($1/C_2$) of this layer with increasing of oxalic acid concentration reflects the strong adsorption of the oxalate anion with increasing of its concentration and increasing of hydrogen evolution. Moreover, the presence of diffusion process at the interfacial layer of the electrode indicates again the formation of porous film and the decreasing of diffusion impedance indicating the increase of electrolyte diffusion through the pores, as a sequence of increasing of the porosity with the increase of oxalate concentration.

At the lowest concentration of oxalic acid (0.01 M) with highest corrosion resistance, the tested electrode was immersed in this solution containing either Cl^-, F^- or PO_4^{3-} ions with various anion concentrations (0.01 to 1.0 M). Figures 10 and 11 show Bode plots as examples for Cl^- and PO_4^{3-} ions, respectively. The EIS results of the tested electrode were analyzed, following the suitable proposed model in Figure 4c.

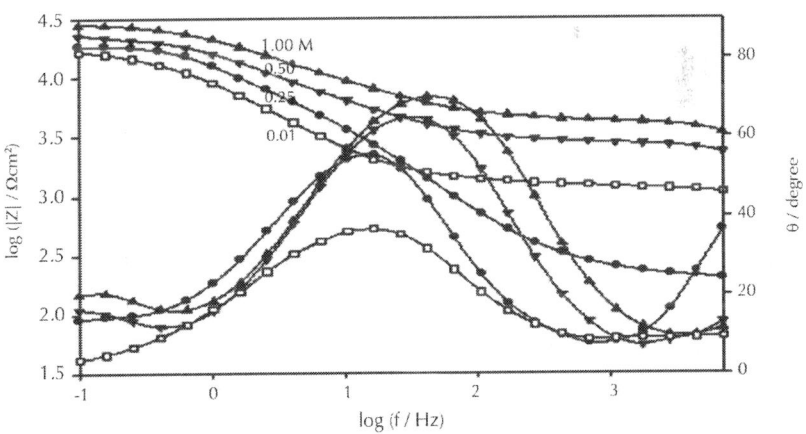

Figure 10: Bode plots of AZ31E electrode as a function of concentration for Cl^- anion in naturally aerated 0.01 M oxalic acid solution, at 298 K.

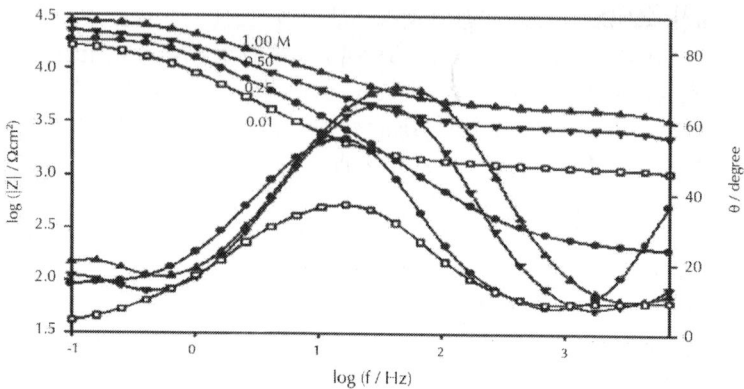

Figure 11: Bode plots of AZ31E electrode as a function of concentration for PO_4^{3-} anion in naturally aerated 0.01 M oxalic acid solution, at 298 K.

The theoretical simulated parameters for the tested alloy at each concentration from the added anions (Cl⁻, F⁻ or PO_4^{3-}) to the forming 0.01 M oxalic acid solution were computed and summarized in Table 5.

Table 5: Impedance and corrosion parameters of AZ31E electrode as a function of concentration for Cl⁻, F⁻ and PO_4^{3-} anions in naturally aerated 0.01 M oxalic acid, at 298 K

Anion	C	R^1	c^1	1	R^2	W	c^2	2	R^s	i_{cor}^r
	M	$K\Omega$ cm^2	μF cm^{-2}		Ω cm^2	kDW	nF cm^{-2}		Ω cm^2	μA cm^{-2}
Cl⁻	blank	2.10	4.7	0.84	58.3	10.2	19.2	0.58	243	31.7
	0.01	20.0	2.5	0.83	94.1	5.30	7.60	0.57	501	13.4
	0.25	14.3	2.9	0.83	87.4	4.30	11.9	0.56	80	16.7
	0.50	10.5	3.9	0.81	82.1	3.10	14.9	0.54	100	20.6
	1.00	6.70	4.1	0.84	63.7	1.50	18.6	0.53	589	25.1
F⁻	blank	2.10	4.7	0.84	58.3	10.2	19.2	0.58	243	31.7
	0.01	23.8	2.1	0.81	110	9.30	5.30	0.59	208	10.8
	0.25	21.3	2.5	0.87	106	8.60	6.40	0.67	80	11.2
	0.50	18.6	2.6	0.87	91.6	7.50	9.60	0.62	100	12.1
	1.00	15.8	2.8	0.81	69.2	6.70	10.1	0.59	589	14.0

PO_4^{3-}	Blank	2.10	4.7	0.84	58.3	10.2	19.2	0.58	243	31.7
	0.01	17.0	2.4	0.87	121	4.20	7.40	0.57	1075	30.0
	0.25	20.1	2.1	0.88	143	5.40	6.10	0.52	199	15.8
	0.50	24.2	1.2	0.85	165	6.20	5.60	0.56	2290	3.50
	1.00	28.9	0.6	0.82	198	8.10	4.70	0.65	3388	2.41

In chloride or fluoride additive solutions, the total resistance (Figure 12), Warburg resistance and $1/C$ decreases with increasing its concentrations. As stated previously, this is due to the deleterious effect of chloride ions [10]. For F-ions, an oxidation reaction occurred in the formation of MgF_2 as follows:

$$Mg+2H_2O \rightarrow Mg(OH)_2+H_2\uparrow \qquad (2)$$

Since $Mg(OH)_2$ was not stable in acidic solution [25], reactions should occur as follows:

$$Mg(OH)_2+2HF \rightarrow MgF_2+2H_2O \qquad (3)$$
$$Mg(OH)_2 \rightarrow MgO+H_2O \qquad (4)$$

The overall reaction occurred as follows:

$$Mg+H_2O+2F^- \rightarrow MgO+MgF_2+H_2\uparrow \qquad (5)$$

The pores in the film should be generated by the hydrogen evolution. These pores might be decreased or filled by the precipitation of MgF_2 particles [25], thus the presence of fluoride ions decreases the corrosion of the tested alloy than the blank (0.01 M oxalic acid). However, depassivation process occurs by increasing fluoride concentration due to breakdown of the formed grained layer of MgF_2 that leads to drastic increase in the surface roughness. Furthermore, in presence of F-ions, aluminum which becomes enriched in the surface can form the soluble complex $(AlF_6)^{3-}$, thereby, participates at higher F-ions concentrations in decreasing the stability of the passive surface film on AZ31E alloy [14]. However, Cl-ion is more strongly adsorbed on the alloy surface than F-ions, so, its resistivity is lower than fluoride ion.

For phosphate anion as additive, increasing the total resistance (R_T) (Figure 12), W and $1/C_1$ with increasing of phosphate concentration indicates that interaction between oxalic acid and phosphate forms phosphate complexes that increase with increasing phosphate concentration and leads to passivation of AZ31E surface. Also, by measuring the pH of the medium, it increases slightly from acidic ~ 6.8 to basic medium reaching to 11.1 at 1.0 M phosphate concentration, leading to passivation. At pH 11.1, HPO_4^{2-} species has almost equal tendency for existing in solution as PO_4^{3-} anions and thus the solution at this pH will contain the two phosphate species with nearly equal relative fraction [2, 26-27]. Therein the electrolyte pH plays a determinant influence on film properties, where films formed in phosphate solutions at higher pH values are thicker of better protection for the alloy than those formed in acidic ones.

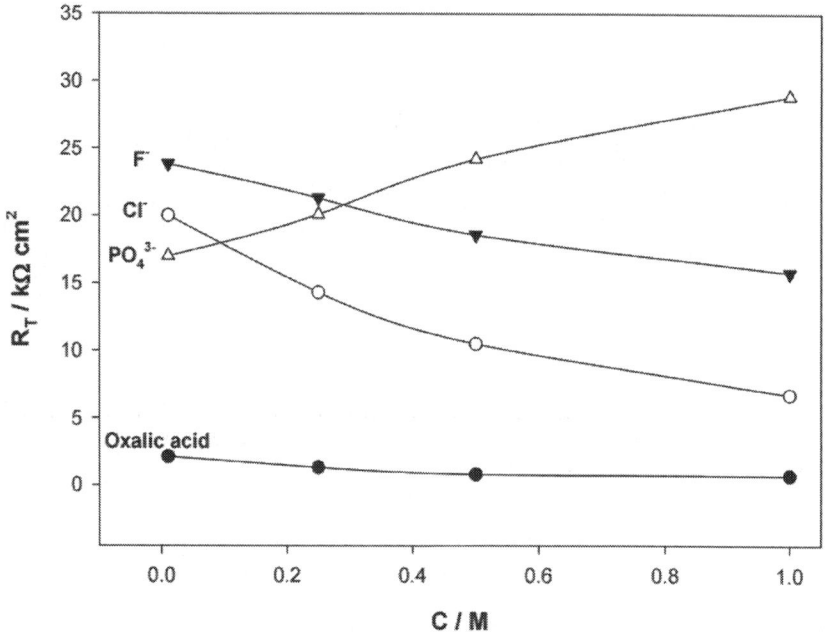

Figure 12: Variation of R_T of AZ31E electrode as a function of concentration for oxalic acid and Cl^-, F^-or PO_4^{3-} anions in naturally aerated 0.01 M oxalic acid solution, at 298 K.

Impedance results are in good agreement with polarization data.

AZ31E Alloy in Oxalic Acid Solution [12]

The Potentiodynamic polarization behavior of the AZ31E electrode was studied in relation to concentration of oxalic acid electrolyte.

Figure 13 shows the scans for the tested electrode in 0.01 M oxalic acid solution with different concentrations (0.01-1.0 M) of PO_4^{3-} ion, at a scan rate of 1 mV/s over the potential range from -2.0 to 1.0 V vs. SCE. Prior to the potential sweep, the electrode was left for 2 hours until a steady state potential was reached.

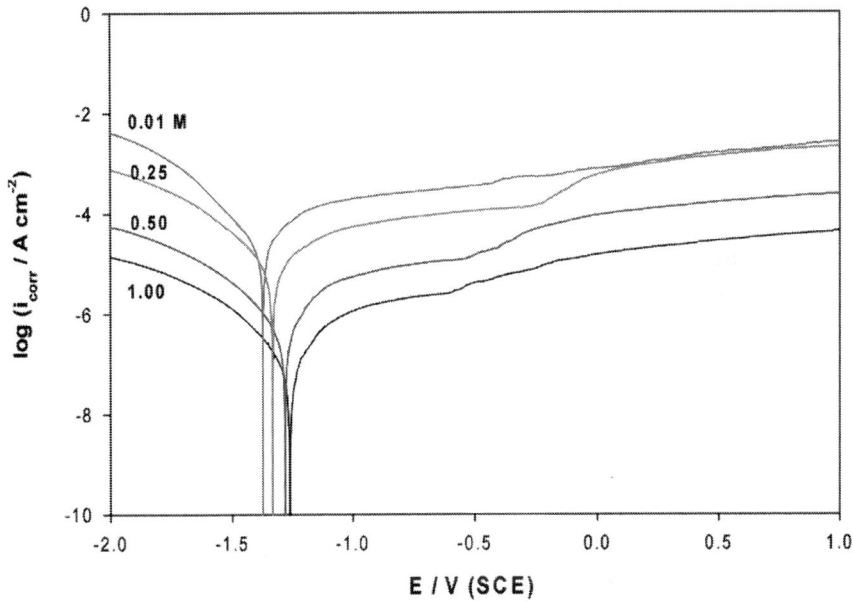

Figure 13: Potentiodynamic polarization scans of AZ31E electrode as a function of concentration for PO_4^{3-} anion in naturally aerated 0.01 M oxalic acid solution, at 298 K.

On increasing the concentration of oxalic acid, an increase in the corrosion current density was observed (Figure 14). This may reflects the changing of the nature of the film formed on the surface (may represents the replacement of MgO by $Mg_2C_2O_4$).

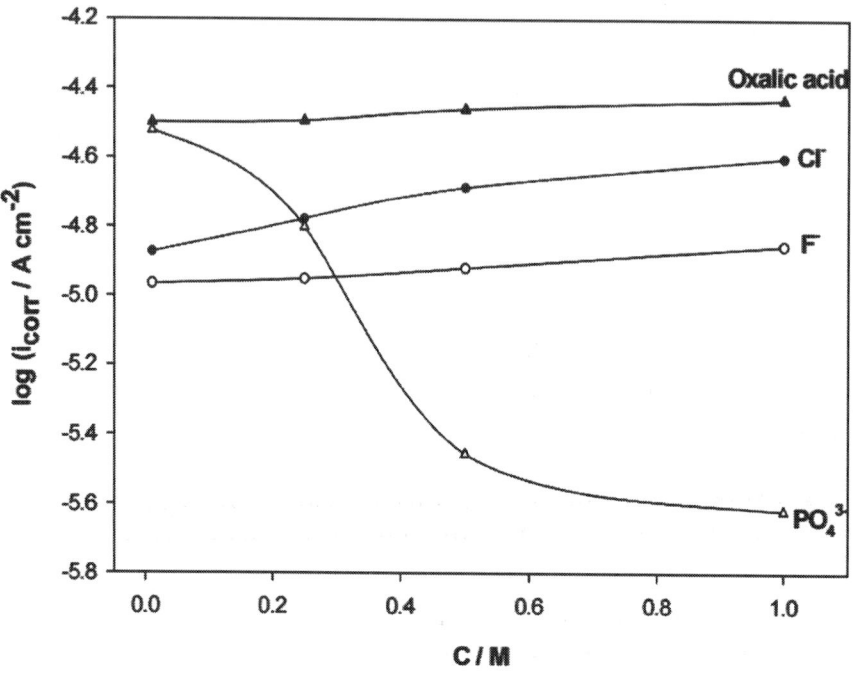

Figure 14: Variation of logarithm of corrosion current density (i_{corr}) for AZ31E electrode as a function of concentration for oxalic acid and Cl^-, F^-or PO_4^{3-} anions in naturally aerated 0.01 M oxalic acid solution, at 298 K.

The effect of added Cl^-or F^-or PO_4^{3-} ions on the electrochemical behavior of the tested AZ31E electrode in 0.01 M $H_2C_2O_4$ solution at 298 K is shown in Figure 15. It was found that i_{corr} value increases with increasing either Cl^-or F^-ion concentration reflecting the harmful effect of both ions leading to an increase in hydrogen evolution [28-29] and corrosion rate. For phosphate ion, i_{corr} value decreases with increasing its concentration that is the corrosion rate decreases. So, phosphate is useful in reducing the corrosion or hydrogen evolution rate [30-31]. These results coincide with that drawn from EIS data.

Generally, for comparing corrosion rate obtained from Tafel and EIS measurements, it is well known that the polarization resistance R_p is related to the corrosion rate through Tafel slopes β_a and β_c by Stern–Geary equation [32]:

$$i_{corr} = \frac{1}{2.303 R_P \left(\dfrac{1}{\beta_a} + \dfrac{1}{|\beta_c|} \right)} \tag{6}$$

Table 6: Corrosion rate (P_i) calculated from EIS and Tafel methods of AZ31E electrode as a function of concentration for Cl$^-$, F$^-$and PO$_4^{3-}$ anions in naturally aerated 0.01 M oxalic acid, at 298 K

Anion	C	R^T	icorr (EIS)	Pi (EIS)	icorr (Tafel)	R^P	Pi (Tafel)
	M	KΩ cm2	µA cm-2	mm/y	µA cm-2	kΩ cm2	mm/y
Cl-	blank	2.10	109.4	2.50	31.7	7.24	0.72
	0.01	20.0	14.28	0.33	13.4	21.3	0.31
	0.25	14.3	19.41	0.45	16.7	16.7	0.38
	0.50	10.5	26.16	0.59	20.6	13.4	0.47
	1.00	6.70	41.1	0.93	25.1	10.9	0.57
F-	blank	2.10	109.43	2.50	31.7	7.24	0.72
	0.01	23.8	18.93	0.43	10.8	41.5	0.25
	0.25	21.3	21.44	0.49	11.2	40.8	0.26
	0.50	18.6	25.57	0.58	12.1	39.5	0.28
	1.00	15.8	27.07	0.62	14.0	30.6	0.32
PO^{43-}	blank	2.10	109.43	2.50	31.7	7.24	0.72
	0.01	17.0	41.98	0.96	30.0	23.8	0.69
	0.25	20.1	32.36	0.74	15.8	41.2	0.36
	0.50	24.2	8.391	0.19	3.50	58.1	0.08
	1.00	28.9	5.883	0.13	2.41	70.8	0.06

As given in Table 6, it can be seen that evaluated R_p values obtained from Tafel measurements have the same trend as R_T obtained from EIS measurements. By calculating i_{corr} from EIS measurements using cathodic, anodic slopes and R_T, it was found that they also have the same trend as that obtained from Tafel measurements. By calculation of corrosion rate, where i_{corr} (mA cm^{-2}) is related to the average corrosion rate in mm/y (P_i) using [4]:

$$P_i = 22.85 i_{corr} \tag{7}$$

It was found that corrosion rate obtained from EIS method is comparable with that obtained from Tafel extrapolation method. Thus there is a good agreement between corrosion rates determined by both techniques.

CONCLUSIONS

The corrosion rate of magnesium alloy in aqueous ethylene glycol depends on the concentration of the solution. A diluted ethylene glycol solution is more corrosive than a concentrated one at room temperature. Ethylene glycol solution containing $Cl^- > 0.05$ M or $F^- < 0.05$ M are more corrosive than the blank (30% ethylene glycol-70% water). However, at concentrations < 0.05 for chloride or > 0.05 M flouride ions, some inhibition effect has been observed. The corrosion of AZ91D alloy in the blank can be effectively inhibited by addition of 0.05 mM paracetamol that reacts with AZ91D alloy and forms a protective film on the surface at this concentration.

The corrosion rate of AZ31E magnesium alloy in oxalic acid solution depends on the concentration of the solution and the additive. A concentrated oxalic acid solution with lowest pH and highest hydrogen evolution is the more corrosive one. Oxalic acid solution of 0.01 M concentration containing Cl^- or F^- are more corrosive with increasing the concentration from 0.01 to 1.0 M for the anions as observed from impedance or polarization techniques. For PO_4^{3-} anion in 0.01 M oxalic acid solution, it acts passivator. The corrosion rate decreases with increasing its concentration.

REFERENCES

1.   Fekry A., Electrochemical Corrosion Behavior of Magnesium Alloys in Biological Solutions, In Czerwinski F. (Ed.), Magnesium Alloys-Corrosion and Surface Treatments. Rijeka: InTech; 2011. p 65-92.

2.   Heakal F, Fekry A, Fatayerji M. Electrochemical behavior of AZ91D magnesium alloy in phosphate medium-Part I. Effect of pH. J. Applied Electrochemistry 2009;39(5) 583-591.

3. Zhang T, Liu X, Shao Y, Meng G, Wang F. Electrochemical noise analysis on the pit corrosion susceptibility of Mg–10Gd–2Y–0.5Zr, AZ91D alloy and pure magnesium using stochastic model. Corrosion Science 2008; 50(12) 3500-3507.

4. Song G, Bowles A, StJohn D. Corrosion resistance of aged die cast magnesium alloy AZ91D. Materials Science and Engineering 2004;A366(1) 74-86.

5. Song G, Atrens A, Dargusch M. Influence of microstructure on the corrosion of diecast AZ91D. Corrosion Science 1998;41(2) 249-273.

6. Fekry A, Fatayerji M. Electrochemical Corrosion Behavior of AZ91D Alloy in Ethylene Glycol. Electrochimica Acta 2009;54(26) 6522-6528.

7. Song G, StJohn D, Corrosion behaviour of magnesium in ethylene glycol Original Research ArticleCorrosion Science 2004 46(6) 1381-1399.

8. Song G. Corrosion and its inhibition of engine block magnesium alloys in coolants, CAST Report 2001081, 2001 (confidential).

9. PCT/IB99/01659, 1999.

10. Fekry A. The influence of chloride and sulphate ions on the corrosion behavior of Ti and Ti-6Al-4V alloy in oxalic acid. Electrochimica Acta 2009;54(12) 3480-3489.

11. Maruthamuthu P, Ashokkumar M. Hydrogen generation using Cu(II)/WO3 and oxalic acid by visible light. International Journal of Hydrogen Energy 1988;13(11) 677-680.

12. Fekry A. Impedance and hydrogen evolution studies on magnesium alloy in oxalic acid solution containing different anions. International journal of Hydrogen Energy 2010; 35(23) 12945-12951.

13. Beccaria A, Bertolotto C. Inhibitory action of 3-trimethoxysilylpropanethiol-1 on copper corrosion in NaCl solutions. Electrochimica Acta 1997; 42(9) 1361-1371.

14. Heakal F, Fekry A, Fatayerji M. Influence of halides on the dissolution and passivation behavior of AZ91D magnesium alloy in aqueous solutions. Electrochimica Acta 2009;54(5) 1545-1557.

15. Macdonald J, In Macdonald J. (Ed.), Emphasizing Solid Materials

and Systems. John Wiley & Sons, 1987.

16. Macdonald D. Reflections on the history of electrochemical impedance spectroscopy. Electrochimica Acta 2006 51(8-9) 1376-1388.

17. Retter U, Widmann A, Siegler K, Kahlert H. On the impedance of potassium nickel(II) hexacyanoferrate(II) composite electrodes—the generalization of the Randles model referring to inhomogeneous electrode materials. Journal of Electroanalytical Chemistry 2003;546, 87-96.

18. Heakal F, Fekry A. Experimental and Theoretical Study of Uracil and Adenine Inhibitors in Sn-Ag Alloy/Nitric Acid Corroding System. Journal of Electrochemical Society 2008;155(11) C534-542.

19. Song G, Atrens A, Wu X, Zhang B. Corrosion behaviour of AZ21, AZ501 and AZ91 in sodium chloride. Corrosion Science 1998;40(10) 1769-1791.

20. Gulbrandsen E, Tafto J, Olsen A. The passive behaviour of Mg in alkaline fluoride solutions. Electrochemical and electron microscopical investigations. Corrosion Science 1993;34(9) 1423-1440.

21. Chao C, Lin L, Macdonald D. Point defect model for anodic passive films, Journal of Electrochemical Society 1981;128(6) 1181-1187.

22. Liu Y, Wang Q, Song Y, Zhang D, Yu S, Zhu X. A study on the corrosion behavior of Ce-modified cast AZ91 magnesium alloy in the presence of sulfate-reducing bacteria. Journal of Alloys and Compounds 2009;473(1-2) 550-556.

23. Jain AK, Acharya NK, Kulshreshtha V, Awasthi K, Singh M, Vijay YK. Study of hydrogen transport through porous aluminum and composite membranes. International Journal of Hydrogen Energy 2008;33(1) 346-349.

24. Fekry A, El-Sherief R. Electrochemical corrosion behavior of Magnesium and Titanium alloys in simulated body fluid. Electrochimica Acta 2009;54(28) 7280-7285.

25. Ma L, Wang P, Cheng H. Hydrogen sorption kinetics of MgH2 catalyzed with titanium compounds. International Journal of

Hydrogen Energy 2010;35(7) 3046-3050.

26. Heakal F, Fekry A, Fatayerji M. Electrochemical behavior of AZ91D magnesium alloy in phosphate medium–Part II. Induced passivation. Journal of Applied Electrochemistry 2009;39(9)1633-1642..

27. Muñoz L, Bergel A, Féron D, Basséguy R. Hydrogen production by electrolysis of a phosphate solution on a stainless steel cathode. International Journal of Hydrogen Energy 2010;35(16) 8561-8568.

28. Uan JY, Cho CY, Liu KT. Generation of hydrogen from magnesium alloy scraps catalyzed by platinum-coated titanium net in NaCl aqueous solution. International Journal of Hydrogen Energy 2007;32(13) 2337-2343.

29. Ameer M, Fekry A. Inhibition effect of newly synthesized heterocyclic organic molecules on corrosion of steel in alkaline medium containing chloride. International Journal of Hydrogen Energy 2010;35(20) 11387-11396.

30. Fekry A, Ameer M. Corrosion inhibition of mild steel in acidic media using newly synthesized heterocyclic organic molecules. International Journal of Hydrogen Energy 2010;35(14) 7641-7651.

31. Azizi O, Jafarian M, Gobal F, Heli H, Mahjani MG. The investigation of the kinetics and mechanism of hydrogen evolution reaction on tin. International Journal of Hydrogen Energy 2007;32(12)1755-1761.

32. Boudjemaa A, Boumaza S, Trari M, Bouarab R, Bouguelia A. Physical and photo-electrochemical characterizations of α-Fe2O3. Application for hydrogen production. International Journal of Hydrogen Energy 2009;34(10) 4268-4274.

Corrosion of Materials in Liquid Magnesium Alloys and Its Prevention

Frank Czerwinski[1]

[1]Canmet MATERIALS, Natural Resources Canada, Hamilton, Ontario, Canada

INTRODUCTION

Magnesium alloys with their unique physical and chemical properties are important candidates for many modern engineering applications. Their density, being the lowest of all structural metals, makes them the primary choice in global attempts aimed at reducing the weight of transportation vehicles. However, magnesium also creates challenges at certain stages of raw alloy melting, fabrication of net-shape components and their service. The first one is caused by very high affinity of magnesium to oxygen, which requires protective atmospheres increasing manufacturing cost and heavily contributing to greenhouse gas emissions [1] [2] [3]. While magnesium exhibits high affinity to oxygen, at temperatures corresponding to semisolid or liquid states it is also highly corrosive towards materials it contacts [4] [5] [6]. This imposes challenges to the selection of materials used to contain, transfer or process molten magnesium during manufacturing operations.

Understanding the reactivity of liquid magnesium with engineering materials to eliminate or at least reduce the progress of corrosion is paramount not only during fabrication processes but also in other unique applications. They include joining of dissimilar materials where magnesium is one part of the joint couple and involves similar liquid/solid interface phenomena [7] [8]. Another example is the liquid battery cell, having two liquid metal electrodes, e.g. magnesium and antimony, separated by a molten salt electrolyte, that self-segregate into three layers based upon density and immiscibility. Such an assembly faces also corrosion issues [9]. During joining by exploring so-called compound casting [10], where two Mg alloys, one in the solid state and another one in the liquid state, are brought together, an interface formation by solid-liquid reaction is the essence of the phenomena leading to a metallurgical bond. Thus, at present, there are still challenges to be addressed to understand the mechanism of corrosion attack, compatibility of materials in respect to liquid magnesium and designing protection methods.

The aim of this chapter is to review fundamental aspects of corrosion in liquid magnesium alloys, assess degradation of selected metallic and ceramic materials in this environment and define methods of corrosion prevention.

LIQUID METAL CORROSION

Liquid metal corrosion is understood as a physical or physicochemical process that follows the formation pattern of metallic alloys. In contrast to corrosion of metals in aqueous solutions, no transfer of electrons is involved. The essential part of the process is dissolution accompanied by formation of a liquid alloy with mixed composition and chemical reactions resulting in creation of intermetallic compounds. Liquid metals are a group of coolants with increasing importance for high-temperature processes and power engineering. Due to great heat-transfer properties, liquid metals are used in the nuclear industry as heat transfer media [11]. Therefore, the majority of operating experience with liquid metal systems has been accumulated within the nuclear industry. The favorable thermo-physical properties of liquid metals allow for high rates of heat removal in comparison to other coolants, e.g., water/steam. Although nuclear reactor engineering with light liquid metal/alloy sodium and sodium-potassium and heavy liquid metal/alloy liquid lead and lead bismuth differ substantially from a liquid magnesium environment, general aspects of corrosion are similar. Common issues show also other applications such as liquid metal spallation targets in elementary particle sources, galvanizing of steel by hot dipping in a molten metal bath of Zn [12], as well as glass production or electronics cooling. Using liquid metals may enable alternative technologies like direct thermal-electric conversion or use of solarhigh temperature heat in chemical processes [13].

Factors Affecting Progress of Corrosion

Three major steps that control the corrosion progress include (i) transport in solid state; (ii) reactions at the solid/liquid interface dominated by dissolution of the solid alloy and its transport into liquid magnesium and (iii) transport of species within the liquid state (Fig. 1). In a hypothetical static system, corrosion would continue until reaching the solubility limit. In the case of an engineering environment, alloying elements and impurities react with atmosphere forming oxides and nitrides that make the process more complex. In particular, the temperature gradient and the concentration gradient make the system dynamic, leading to mass transfer.

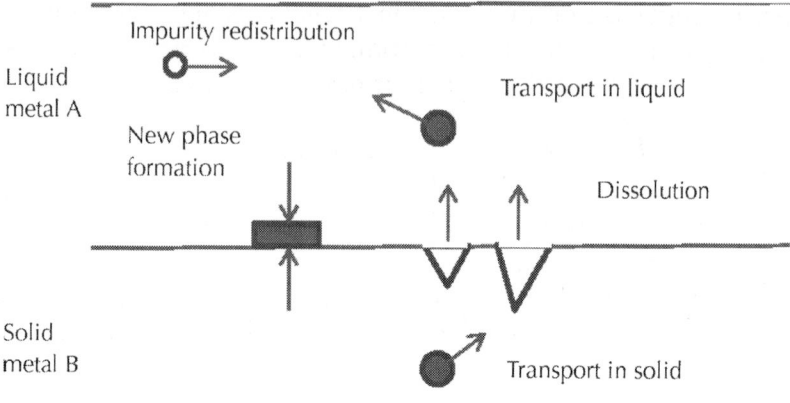

Figure 1: Schematics emphasizing phenomena involved during corrosion in liquid metals.

According to early research [14], the corrosive attacks are classified as: (i) simple dissolution, (ii) alloying between liquid metal and solid metal, (iii) intergranular penetration, (iv) impurity reactions, (v) temperature gradient mass transfer, (vi) concentration-gradient mass transfer or dissimilar-metal mass transfer. As variables controlling liquid metal corrosion, such factors as temperature, its gradient or cyclic fluctuation, surface area to volume ratio, metal purity, flow velocity and some characteristics of material being in contact with liquid metal, are named.

For certain liquid metals, the progress of corrosion may be reduced by adding to the liquid alloy either metallic or non-metallic corrosion inhibitors. It is believed that inhibitors decrease the corrosion rate by forming a protective film, separating the metallic surface from corrosive media. For example, metallic inhibitors added to liquid Pb and Pb-Bi caused the formation on a steel surface of a protective layer of nitrides or carbides by the chemical reaction between inhibitors and carbon and nitrogen in steel [11]. In particular, for carbon steels and low-alloy steels, additions of Zr and Ti were confirmed to be effective. At the same time, the effectiveness of Zr and Ti for stainless steel, being in contact with liquid Pb and Pb-Bi alloys, was very low.

An essential factor in corrosion progress is the corrosion front morphology. The ideally uniform corrosion front is possible in theory when the diffusion rate in the solid is fast enough to balance the mass transfer rate in the liquid. In practice, due to the fact that diffusion in solid is the slowest step, there is a development of the front morphology. In single-phase materials, the grain orientation, grain boundaries and impurities will contribute to different rates of corrosion progress over the surface. In multi-phase materials, phase chemistry and their crystallography impose additional complexity. A particular case is the preferred penetration of liquid metal along grain boundaries of the adjacent solid material, causing its embrittlement.

Models of Liquid Metal Corrosion

At early stages of corrosion, the dissolution reaction is considered to be the fastest step, with mass transport in the liquid seen as the controlling step of the entire process. In contrast, at the steady state, either dissolution at the solid/liquid interface or mass transport in the bulk liquid is in control of the corrosion progress. It was revealed during experiments between austenitic stainless steel AISI 316L and liquid Ga, along with its alloys (Ga-14Sn-6Zn, Ga-8Sn-6Zn), the corrosion progress gravimetrically measured by metal losses, was significantly lower in Ga alloys than in pure Ga [15]. As an explanation, the lower diffusivity of species in pure liquid Ga than in Ga alloys was proposed.

For a nuclear reactor coolant system, a number of models were developed allowing the calculation of the corrosion rate and corrosion layer thickness [16]. They explore a mathematical analysis of the transport of various species in the solid phase, in flowing liquid phase and mass exchange at the interface between solid and liquid. The negative side of modelling so far is an exclusion of the pitting corrosion, stress corrosion cracking and liquid metal embrittlement due to a lack of theoretical and experimental data.

Liquid Metal Embrittlement

While being in contact with liquid metal under stress, certain metals experience a drastic reduction in ductility [17] [18]. It should be pointed out that the loss of ductility takes place in normally ductile metals.

Often the phenomenon is seen as the crack propagation associated with a change of the fracture surface from ductile into brittle of an intergranular type. According to the conventional mechanism, the lower melting point liquid metal fills a crack in the solid metal, thereby weakening material at the crack tip and allowing it to propagate at much lower stress. Moreover, the traditional models considered the grain boundary segregation associated with formation of the interfacial phase. Existing failure mechanisms, including the decohesion model, adsorption induced dislocation emission model and dissolution condensation mechanism model are shown schematically in Fig. 2 [19]. There are many combinations of the matrix and embrittling solute [20], including steel embrittlement by liquid Cu, aluminum by Hg, or stainless steel by liquid Zn. The literature does not report that liquid magnesium is the embrittling solute but some alloying elements of magnesium alloys, e.g. Zn, are known as causing embrittlement.

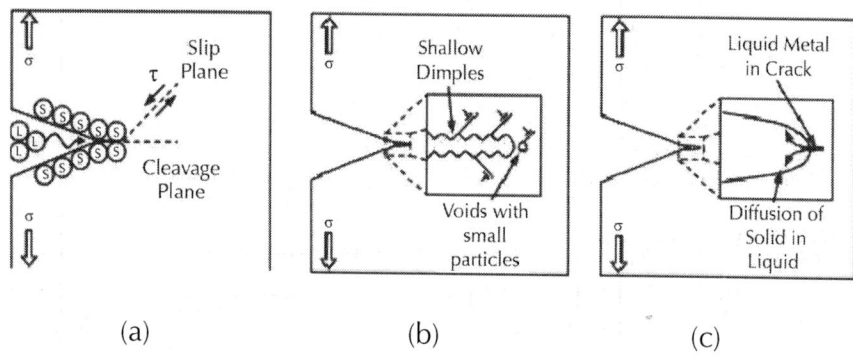

(a)   (b)   (c)

Figure 2: Existing failure mechanisms of corrosion in liquid metals: (a) decohesion model; (b) adsorption induced dislocation emission model; (c) dissolution condensation mechanism model [19].

There are also practical observations, which point out that: (i) liquid metal must be present and (ii) the area affected must be under stress [21]. According to Ref. [22], the embrittlement will occur even when the liquid metal is removed from the solid metal surface before stressing. As factors controlling the material susceptibility to liquid metal embrittlement, the low mutual solubility between the liquid and solid metals and an absence of intermetallic compound formation between the solid-liquid couple are frequently quoted. As revealed by a number of advanced analytical techniques of surface investigation, the micro-mechanism of liquid metal corrosion is more complex. For example, in the case of nickel infused with bismuth atoms a bilayer interfacial phase was detected by aberration-corrected scanning transmission electron microscopy [23]. This observation showed that adsorption can induce a coupled grain boundary structural and chemical phase transition leading to embrittlement. Experiments with several reactive couples including liquid Ga into Al, Hg into Ag and Ag into Cu showed that the size of the brittle fracture surface area is proportional to logarithm of the exposure time. There is also evidence [22] of the existence of an incubation period, being inversely proportional to the penetration speed of the liquid metal into solid metal. During a study of the 9%Cr martensitic steel in stagnant liquid environments of Pb, Pb-Bi or Sn the steel embrittlement can be explained by a reduction of the surface energy of the bare metal induced by some adsorption of the liquid metal [24]. This fact allowed understanding some experimental observations including (i) instantaneous effect of embrittlement with no requirements of solid-state diffusion; (ii) reduction of embrittlement with increasing temperature; increased propagation of surface crack into bulk alloy caused by cyclic loading.

Ceramic Materials in Contact with Liquid Metals

In contrast to metallic materials, ceramics show generally higher resistance to corrosive attack at high temperatures [25]. Due to degradation of mechanical properties of metallic materials, at a certain temperature range, ceramics represent the only choice for many advanced engineering designs. This applies also to an environment of liquid metals. In the case of ceramics being in contact with molten metals, the destructive process involves a reduction-based removal of non-metallic elements from the solid, undermining their structural integrity [26] [27]. It is also claimed that ceramics when in contact with liquid low-melting metals experience the strength reduction. According to experiments with A-995 alumina exposed to liquid Sb, Pb, Ca, Bi and 50Bi-30Pb-20Sn*alloy (* all alloy compositions are in weight % unless indicated otherwise), a considerable reduction in strength by alumina was experienced [28]. In all cases, alumina ceramic was wetted by liquid metal, which adhered to their surface after solidification.

CORROSIVE NATURE OF LIQUID MAGNESIUM AND ITS ALLOYS

In engineering practice, to generate the sufficient strength and other mechanical properties, magnesium is alloyed with other metals such as Al, Zn, Mn, Si, Cd, Ag, Zr, Ca, Sr, Be or rare earths. Moreover, magnesium alloys contain difficult to remove impurities, mainly Fe, Ni or Cu. Both the alloying elements and impurities change the corrosivity of liquid Mg in regards to materials they contact.

Corrosivity of Pure Magnesium

A selection of materials to sustain liquid magnesium, in terms of resistance to its corrosive attack and mechanical properties degradation due to high temperatures is a key task for engineering applications. It is generally known that iron is inert to molten magnesium. According to the Mg-Fe equilibrium phase diagram, below 1000 °C, Mg does not dissolve in Fe. In a Mg-Fe system, some solubility of Mg is possible only at high pressures and high temperatures. As reported in Ref. [29], the maximum solid solubility of Mg in Ni is 0.00043 at. % and the eutectic composition is at 0.008 at. % Fe.

In contrast to Fe, liquid magnesium reacts with nickel. According to Ref. [30], the solid solubility of Mg in Ni is less than 0.2. at. % Mg at 500 °C. In an Mg-Ni system there is one peritectic and two eutectic reactions [31]. There are two intermetallic compounds with Mg_2Ni melting at 760 °C and $MgNi_2$ melting at 1147 °C [32]. As pointed out in [33], the $MgNi_2$ phase extends from 66.2 to 67.3 at. % Ni.

During analysis of the liquid Mg-solid Ni reaction couple it was noticed that the compounds which are at the Mg/Ni interface, rich in the lower melting point elements, are the first phases formed in metal/metal binary reactions, i. e., Mg_2Ni [34]. It is quite surprising to find that the minority elements, such as Cu or Ni impurities, are the dominant diffusing species. This fact differs from results obtained in many metal/metal systems where the major diffusion species are usually the majority elements in the initially formed compounds. During an experiment with a diffusion couple of 99.999% purity liquid Mg at a temperature of 660 – 680 °C and solid Ni plate the intermetallic phase of Mg_2Ni

grew at the interface and the Ni layer dissolved by diffusion throughout the intermetallic phase into liquid Mg [35]. Although according to the equilibrium phase diagram both phases Mg_2Ni and $MgNi_2$ could form, only the former one was detected.

Effect of Major Alloying Elements

The number of elements that can be added during alloying is rather low due to their limited solubility in magnesium and a competition from phases formed between additions themselves. Of all alloying elements, aluminum is the most often used and its content, reaching up to 11%, is the largest of all metals explored for that purpose. Under equilibrium conditions the solubility of Al in α-Mg is 12.7% at 437 °C but in as-cast alloys a solid solution below 437 °C is enriched only with 2 – 3% of Al [36]. In general, Al is not distributed uniformly and in addition to macro-segregation, also micro-segregation occurs and is related to formation of precipitates and grain boundary enrichment.

The role of aluminum in liquid metal corrosion is substantial due to its large concentration in many Mg alloys and extremely high corrosivity [37]. An example showing Al reactivity when Mg-10%Al alloys contacts carbon steel is shown in Fig. 3. Molten aluminum dissolves practically all conventional alloys including, Fe, Ni, Cr and Co based grades. Although some alloys, such as Ni-Cr-Fe are used with molten Al, their lifespan is rather short. There is some potential of Ti alloys where protection against molten Al is believed to be provided by Ti oxides forming on Ti surface, thus separating both metals. The literature emphasizes also a strong influence of small amounts of Al present in Mg alloys. For example, less than 0.002 wt. % of Al or Si in liquid Mg at 723 °C reacts with Fe from steel creating α-Fe (Al, Si) solid solution [38]. By combining carbon from steel, impurities present in Mg form a compound of Fe_2 (Al,Mg)C. It has been concluded that the above phases were formed not only by solid-state diffusion in steel but also by dissolution-precipitation processes and migration of Fe in liquid magnesium. During a contact of molten magnesium alloys with H13 steel Al content was found to be the controlling factor in the type of phases formed [39]. For Al contents below 6%, FeAl/ (Fe, Mn) Al was the major phase formed with traces of $FeAl_2$ and Fe_2Al_5. For higher Al contents, the Fe_2Al_5 phase dominated and for Al exceeding 12%, the $Fe_{14}Al_{86}$ was the dominant one.

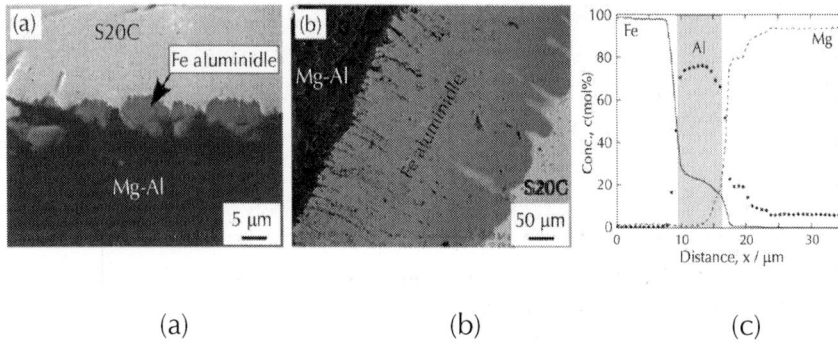

(a)   (b)   (c)

Figure 3: Microstructure of the interface between Mg-10% Al alloy and S20C carbon steel after short term (a) and long term (b) exposures. The concentration profile across interface (a) is shown in (c) [82].

A common ingredient of magnesium alloys is also zinc with a maximum solubility of 8.4%. Additions of zinc are used often in a combination with aluminum to improve the alloy strength at room temperatures or with zirconium and rare earths to generate precipitates that are stable at increased temperatures. From the shape of the αMg-β boundary at the equilibrium phase diagram of Mg-Al-Zn [40], it is seen that the addition of Zn allows more β precipitates to form at a fixed Al content of 8%. The β compound has a part of Al atoms substituted by Zn. Since for Mg-9Al-1Zn alloy at temperatures below 437°C it has the form of $Mg_{17}(Al,Zn)_{12}$, likely $Mg_{17}Al_{11.5}Zn_{0.5}$, the Zn content is expected to be higher [41].

Zinc, which melts at 419°C is a commonly used low melting metal with vast data and broad application experience. Both molten zinc and its alloys are used for hot-dip galvanizing of steel. There is solid evidence that they may cause liquid metal embrittlement of steel at temperatures as low as 400 °C. For example, liquid metal embrittlement can be induced into austenitic stainless steels by molten Zn during its welding with galvanized carbon steel [42]. It is known that refractory metals like Mo, W and their alloys withstand the corrosive attack by molten Zn up to 500 °C. During an experiment with testing Mo, Mo-30%W, W, DIN 1.4841 steel and graphite in a molten Zn environment at 500, 600, 650 and 700 °C for 168 h, differences in corrosion behavior were recorded [43]. While steel was dissolved, Mo was only partly attacked at 500 and 650 °C. At the same time, Mo-30%W, pure W and graphite showed satisfactory corrosion resistance.

Effect of Minor Elements and Impurities

As emphasized already in section 3.2, the chemical compatibility of molten magnesium towards other materials may be altered by the presence of small amounts of other elements. They are not only limited to difficult-to-remove impurities but also to elements deliberately added in small quantities. Many magnesium alloys contain manganese; its major role is designed to bind harmful impurities of iron and heavy metals into harmless intermetallic compounds. In general, Mn has very limited solid solubility, reaching 2.2%, which in the presence of aluminum is further reduced to about 0.3% due to formation of MnAl, $MnAl_6$ and $MnAl_4$ compounds. Its maximum content in commercial magnesium alloys does not exceed 1.5 – 2 %. Based on free energy of oxide formations, Mn looks to be much less reactive than Mg or Al. According to observations in Ref. [44], in liquid Al alloys, a presence of Mn increased oxide layer thickness. Although oxide layer thickness increased with Mn content in Al alloy, oxide contained very small amounts of Mn. It was noticed during industrial practice that e.g. soldering is more likely to occur when die casting magnesium alloys such as AM60 with higher manganese content are used [39]. The die casting trials were also carried out to confirm the soldering development in high pressure die casting conditions [45]. The results showed that the formation of intermetallics started with the nucleation of the η-Fe_2Al_5 phase. During the next stage, manganese substituted some of iron and this phase became $(Fe,Mn)_2Al_5$. Finally, a metastable phase $Mn_{23}Al_{77}$ was formed at the outer layer of the surface exposed to liquid magnesium alloy.

During reaction of solid mild steel and liquid Mg-Mn alloys at 727 °C the chemical interaction was found to depend on the Mn content [46]. In all cases, the interface reactions led to formation of phases from the Fe-Mn phase diagram with the mechanism being dominated by solid state volume diffusion of Fe and Mn. The intergranular melt infiltration was added as the secondary phenomena As shown in Fig. 4, for 0.6 – 0.7 at %Mn, two sublayers were formed with chemistry of αFe (Mn) and γFe (Mn) [46]. For higher content of 1.3 at %Mn, corresponding to the Mn saturated solution, the layer contained βMn (Fe).

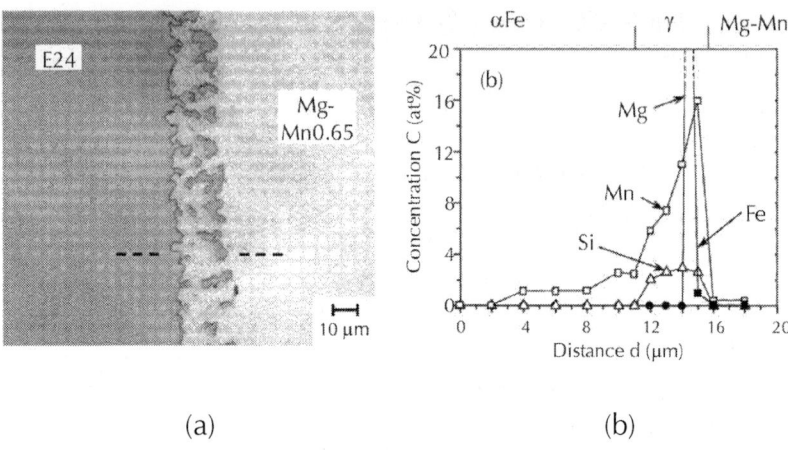

(a)   (b)

Figure 4: Interface between E24 mild steel and Mg-0.65 at% Mn liquid alloy after 65 h exposure at 727 °C (a) along with concentration profile across the interface (b) [46].

The reactivity of liquid Mg-Si alloys containing up to 3.1 at % Si with mild steel at temperature between 677 and 727 °C led to formation of a continuous layer at the solid/liquid interface [47]. For very low Si content of 0.025 at. % the layer consisted of αFe solid solution with less than 2 at % Si For higher Si contents from 0.045 at. % to 3.1 at % the reaction layer consisted of $\alpha_1 Fe_3 Si$ ordered phase with Si content from 24.5 to 27 at. %. The reaction phase was formed by the solid-phase volume diffusion. Morphology of the solid/liquid interface and accompanying concentration profiles are shown in Fig. 5.

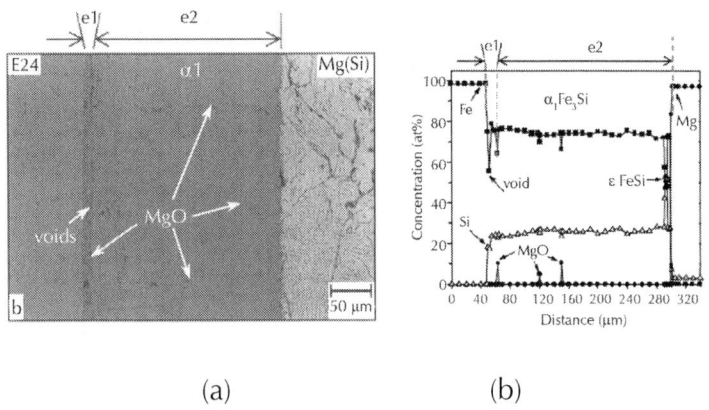

(a)   (b)

Figure 5: Interface between E24 mild steel and liquid Mg-3 at% Si after 250 min at 727 °C (a) and concentration profile showing location of Si in the reaction zone (b) [47].

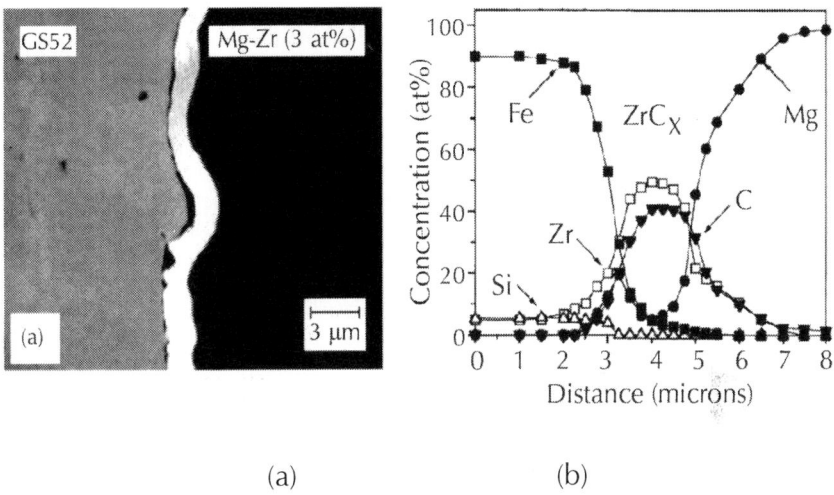

(a)   (b)

Figure 6: Interface between cast iron and Mg-3 at% Zr alloy after exposure of 65 h at 727 °C (a) and concentration profile showing Zr location in the reaction zone (b) [48].

The role of transition alloying elements in Mg alloys on their corrosivity in liquid state was also studied. It includes, for example the reaction between liquid Mg-Zr alloy at 727 °C and pure Fe, steel and cast iron [48]. Although iron is chemically compatible with liquid magnesium, it reacted with Mg-Zr alloys (Fig. 6). The primary reaction took place between Zr in liquid Mg-Zr alloys and iron, forming in the liquid the Fe-Zr compounds, mainly Fe_2Zr. The reaction mechanism is based on dissolution of the solid steel and crystallization within the Mg-Zr liquid alloy, enriched by the solute. The secondary reaction is between Zr and C from the steel substrate to form ZrC_x as a surface layer inhibiting corrosion progress.

COMPLEX NATURE OF MATERIAL DEGRADATION BY LIQUID MAGNESIUM IN ENGINEERING APPLICATIONS

The materials in contact with liquid magnesium during manufacturing operations are subjected to multiple deteriorating effects which accompany the corrosive attack by molten metal. In addition to the corrosive attack of a purely chemical nature, there is an influence of accompanying heat and stress. Depending on processing details, the relative proportion of individual deteriorating factors and their contribution to overall damage experienced by materials are different.

Simultaneous Effect of Stress and Corrosion

The components of magnesium processing equipment in direct contact with liquid metal are a subject of very demanding requirements in terms of materials used. In some cases, tight tolerances may not allow for substantial size change due to wear or corrosion during service. Similarly, high strength requirements may not allow for a reduction in mechanical properties over service time. The most common mechanisms of material degradation in liquid magnesium include [1]: (i) high temperature fatigue; (ii) thermal fatigue; (iii) corrosion fatigue, (iv) creep and stress rupture and/or (v) oxidation. The relative contribution of each mechanism depends on the specific application. An example of a combination of liquid metal, corrosion and fatigue (corrosion fatigue) is shown in Fig. 7. Under a cyclic load caused by melt pressure, corrosion sites act as stress risers, causing progressive crack propagation and premature failure.

Surface
penetrated by
liquid magnesium

Figure 7: Fracture surface of tool steel after service in an environment of liquid magnesium alloy.

Stress due to Difference in Thermal Expansion

At relatively high service temperature, materials experience large dimensional changes due to thermal expansion. As may be deduced from Table 1, for components with a length of 2 m and service temperatures over 620 °C, an elongation may reach several millimeters. Such a difference should be considered during material selection. The key challenge occurs when two or more different materials are in direct contact at high temperatures. For example, when tool steel components are connected with each other by using Ni based alloy Inconel. As seen in Table 1, the coefficient of thermal expansion of the bolts is larger than that for the steel. To maintain a bolt preload, it is necessary to re-torque the bolts once the component reaches the service temperature. Conversely, when the assembly is subsequently cooled to ambient temperature, bolts will shrink more than the steel component.

Hence, without loosening of fasteners, this will cause large tensile stress to develop and fastener stretching that may eventually lead to premature failure. Similarly, materials of a die of a high pressure die casting machine or molten alloy distribution systems should exhibit similar thermal expansion or differences should be incorporated into design.

Table 1: Example showing differences in coefficients of thermal expansion for selected alloys applicable in processing liquid magnesium (10^{-6} m/ (m K) Temperature range from 20 °C to the value indicated [86] [87]

Alloy	100	200	300	400	500	600	700	900	1000 [°C]
AISI H13	11. 9	12. 4	12. 3	12. 7	13. 0	13. 3	13. 5		
DIN 1. 2888	9. 9	10. 4	10. 9	11. 3	11. 6	11. 8	11. 9		
Inconel 718	12. 8	13. 6	13. 9	14. 4	14. 8	15. 1			
Stellite 6	11. 3	12. 9	13. 6	13. 9	14. 2	14. 5	14. 7	15. 5	17. 5
Stellite 12	11. 5	12. 1	12. 6	12. 9	13. 3	13. 8	14. 3	15. 2	15. 6
Stellite 21	11. 0	11. 2	12. 0	12. 6	13. 1	13. 6	14. 3	15. 21	

Stress due to Low Thermal Conductivity

Heat that is required for high temperature processing is not applied directly to magnesium. Instead, it is provided entirely from an external source and transferred through walls of a furnace crucible or the sleeve of a transfer pump. To achieve the fast rate of heat transfer, thermal conductivity of material used is of key importance. If a material has low thermal conductivity it not only requires a longer time to melt magnesium but creates a steep temperature gradient across the component wall thereby generating thermal stress. For thick walls the stress build-up may lead to thermal shock failure. It is seen in Table 2, that Ni-base and Co-base alloys with high strength at temperatures of liquid magnesium processing have rather low thermal conductivities. A particular case has a place when a material is in intermittent contact with liquid magnesium as with a nozzle, die/mold or elements of the melt transfer system. If a surface is subjected to frequent heating/cooling cycles, e.g. every 1-2 min, surface fatigue leads to crack formation. A top view of a steel surface damaged by intermittent contact with liquid magnesium is shown in Fig. 8. Surface microcracks, formed as a result of thermal fatigue, are cyclically filled with liquid alloy.

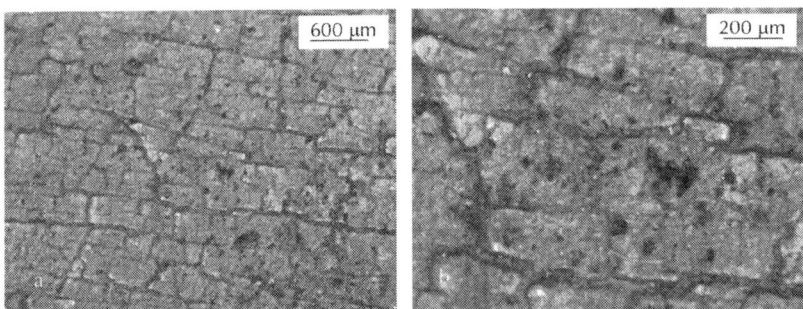

Figure 8: Top view showing surface of tool steel after intermittent exposures to liquid magnesium alloys.

Table 2: Example showing differences in coefficients of thermal conductivity for selected alloys applicable in processing liquid magnesium (W/m °C) [86] [87] [88]

Alloy	20	200	300	400	500	600	700	800	1000 [°C]
AISI H13	25. 5	27. 1		27. 7					
DIN 1. 2888	20. 5	24. 2		27. 5					
Inconel 718	11. 1	14. 1	16. 0	17. 7	18. 8	19. 9	22. 1	23. 7	
Stellite 6	14. 82								
Stellite 12	14. 6								
Stellite 21	14. 5								

MATERIAL SOLUTIONS IN MELTING AND CASTING EQUIPMENT

The industrial equipment for magnesium processing, with components entirely or partly exposed to liquid metal, include crucibles of melting furnaces, elements of die casting and other machinery, parts of pumps and transfer systems. Although the primary requirement imposed on materials used includes the chemical resistance to molten magnesium, there are also other essential properties needed, depending on specific service conditions such as a sustainable level of strength and toughness, creep resistance as well as the resistance to oxidation in air at high temperatures. There are a number of commercial alloys applicable for this purpose that differ in chemical composition, as well as physical and mechanical properties. Examples of most common industrial solutions are listed in Table 3.

Table 3: Metallic alloys applicable for components of magnesium processing equipment exposed to liquid magnesium alloys

Resistance against liquid magnesium only		Resistance against liquid magnesium accompanied by high temperature strength and wear resistance	
Group of alloys	**Grade examples**	**Group of alloys**	**Grade examples**
Majority of tool steels	excluding (AISI A8, A9, A10)	Highly alloyed special steels	DIN 1. 2888 DIN 1. 2886
Low alloy steels	AISI 1330-41615115-6150	Chromium hot work steels	AISI H10-H19
Ferritic stainless steels	AISI 405, 430, 444	Tungsten hot work steels	AISI H21-H26
Martensitic stainless steels	AISI 403, 410, 440		
Special alloys	Nb-30Ti-30W		

Crucibles and Melting Furnaces

The magnesium melting furnaces utilize as a heat source the electric resistance or gas heaters. Thus, an external surface of the crucible is additionally degraded by an exposure to the heat source. Because carbon steel scales on the outside of the melting crucible, being in contact with flame, other material combinations have been tried. The present furnace crucibles are manufactured of (i) ferritic stainless steel; (ii) bimetallic materials with an interior made of low carbon steel and an exterior of stainless steel; (iii) bimetallic design where an interior of low carbon steel is protected from the outside by nickel alloys, forming the oxidation resistant surface adjacent to the furnace heat source. An example of crucibles for handling liquid magnesium is shown in Fig. 9.

Figure 9: Commercial processes requiring equipment resistant to liquid magnesium: (a, b, c) crucibles for magnesium melting [83].

Magnesium crucibles for operation up to 850 °C could be manufactured of wrought or cast mild carbon steels with negligible nickel content. Welding is the dominant joining technique. In the case of using just carbon steel, a crucible lifetime is generally short, sometimes of the order of several days. An improved design of the single material crucible utilizes a ferritic stainless steel, such as AISI 444 [49] [50]. The steel is welded using a nickel-free and high chromium stainless rod. As a result, the crucible lifetime is increased from six to ten times. Since ferritic stainless steels exhibit a tendency to brittle cracking, as a further improvement, bimetallic crucibles were introduced. During melting at 650°C, magnesium tends to leach the Ni out of Ni-containing alloys; a thick austenitic steel base such as AISI 316 is overlaid with a mild carbon steel liner such as AISI 1005 to protect it from a corrosive attack [51]. Another design suggests mild steel crucibles or a lining of 430 stainless grades. The nickel-chromium-iron alloy outside provides high temperature strength and oxidation resistance while the carbon steel inside is more compatible with the molten magnesium [52]. An example of reactivity of Mg with a crucible and a melt transfer pump is shown in Fig. 10.

Figure 10: Magnesium reactivity with (a) crucible holding liquid magnesium alloy; (b) element of transfer system for liquid magnesium alloy with surface covered by solidified alloy.

Die Casting Machines

There are two techniques of die casting and corresponding machines for both processes with essentially different exposure conditions of their components to liquid magnesium. As a result, both applications require quite different material solutions.

During cold-chamber die casting, shown schematically in Fig. 11, the alloy melting is performed in a separate furnace. Then, a portion of molten material is transferred from the furnace crucible into the machine shot sleeve, where a hydraulically operated plunger pushes the metal into the die. The amount of liquid alloy transferred to the sleeve is larger than the part volume so the extra material is used to apply pressure during solidification in the die cavity, thus reducing the generated shrinkage. The typical injection pressure of a cold chamber system exceeds 70,000 kPa. Due to a need to transfer the molten metal from the furnace to the cold-chamber machine, the cycle time is reduced depending on the melt transfer solution.

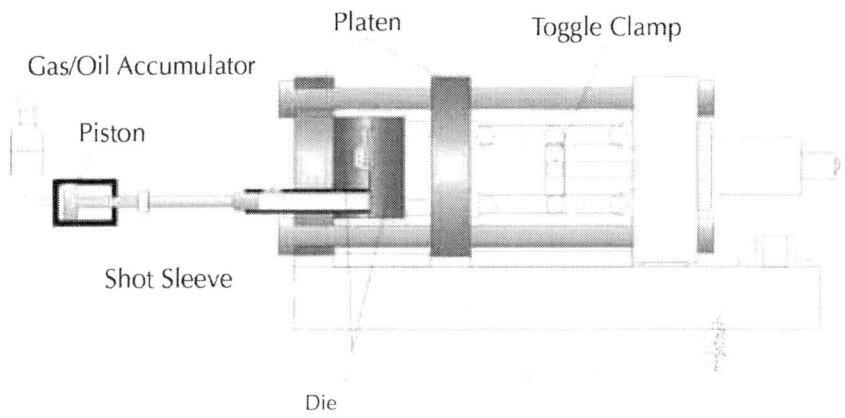

Figure 11: Schematics of cold chamber die casting machine with major components exposed to liquid magnesium alloys [84].

There are a number of components within the machine that are exposed to liquid magnesium, including shot sleeve, plunger, seal rings of the plunger, nozzle, sprue etc. An example of the shot sleeve is shown in Fig. 12a. The shot sleeve walls are in an intimate contact with liquid magnesium for only 1-2 seconds or 5% of the entire cycle duration so its average temperature is well below the melting point of magnesium alloy. However, the non-uniformity in temperature distribution as suggested in Fig 12bcauses non-uniform deformation with a tendency to bending. As a result, a piston cannot slide smoothly inside the shot sleeve, causing friction and wear on both components. During flow into shot sleeve, the liquid alloy heats the inner wall surface reaching the highest temperature within the entire component. In extreme cases the sleeve deformation may lead to plunger seizing. A combination of high temperature, chemical attack, along with abrasion imposed by flowing alloy, contributes to the degradation observed. A similar wear location is also reported for short sleeves used in die casting of Al alloys. A frequent location of the sleeve deterioration area is shown in Fig. 12c.

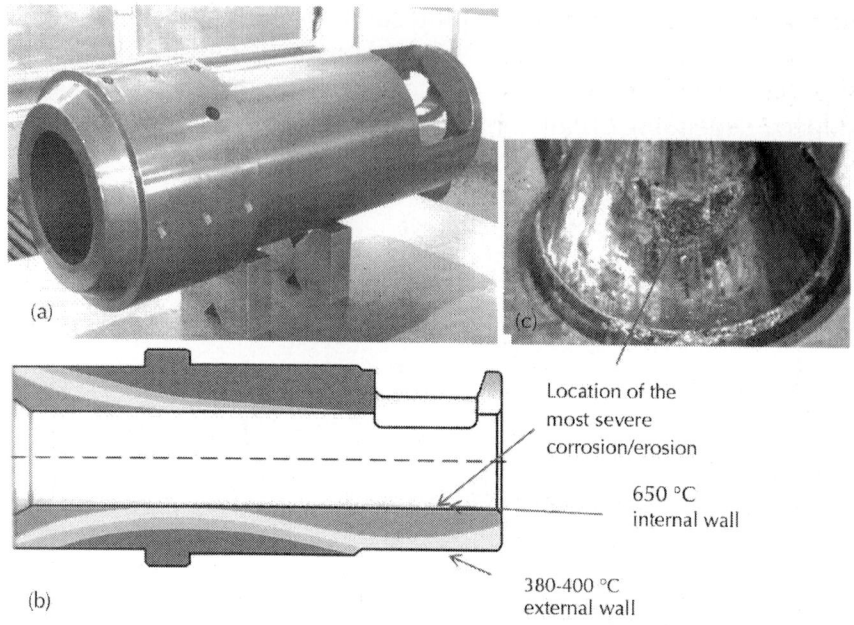

Location of the
most severe
corrosion/erosion

650 °C
internal wall

380-400 °C
external wall

Figure 12: Design of a shot sleeve of a cold chamber die casting machine (a), schematics of temperature distribution (b) and surface deterioration due to a contact with liquid metal (c) [85].

To improve the temperature uniformity within the shot sleeve, thermoregulation is used, employing sleeve channels with circulating cooling/heating media (Fig. 13). The cooling channels help to take away the heat brought by molten alloy. As shown in Fig. 14, this solution substantially lowers the temperature difference between sleeve locations, reducing the extent of thermal fatigue. During selection of material for the shot sleeve there is a trade between durability and cost. As a result, the hot work tool steel of AISI H13 grade performs satisfactorily there.

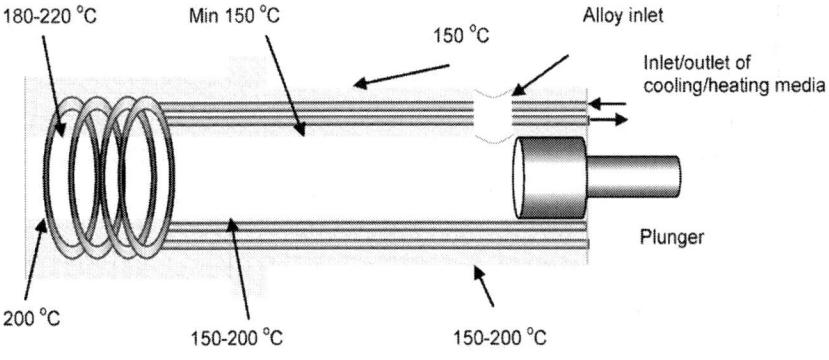

Figure 13: Schematics of thermoregulation within shot sleeve of cold chamber die casting machine [85].

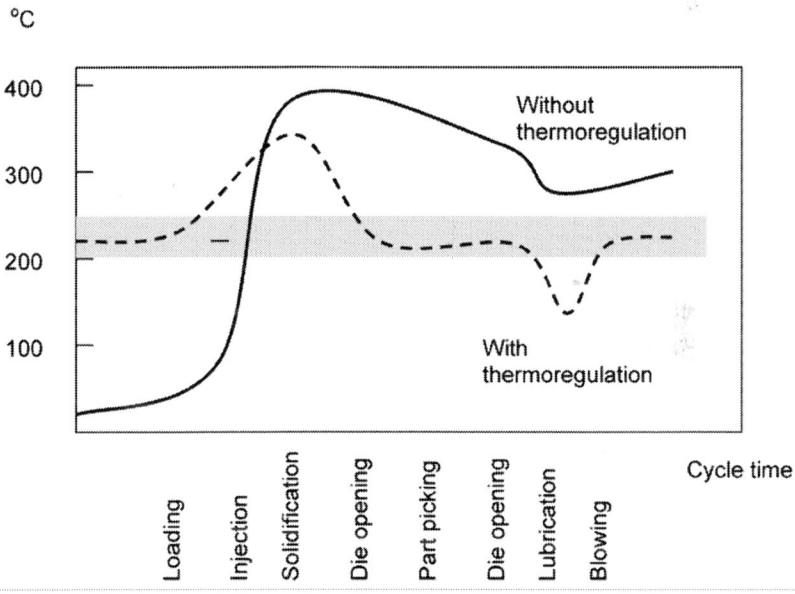

Figure 14: Temperature distribution profile within shot sleeve of cold chamber die casting machine with and without thermoregulation [85].

The hot chamber magnesium die-casting process uses a gooseneck and piston to inject molten magnesium into a die. As shown schematically in Fig. 15, the gooseneck is submerged into molten magnesium so through the side fill holes its cavity is filled with molten magnesium. The piston then forces the molten metal down from the top, delivering it through the gooseneck vertical delivery hole, nozzle and into the die without exposure to the environment. After the metal turns solid inside the die cavity, the die opens, the part is ejected and at the same time the piston retracts to its initial position above the fill holes. Then again, the die is closed, and the gooseneck is filled with molten metal and ready for the next injection cycle.

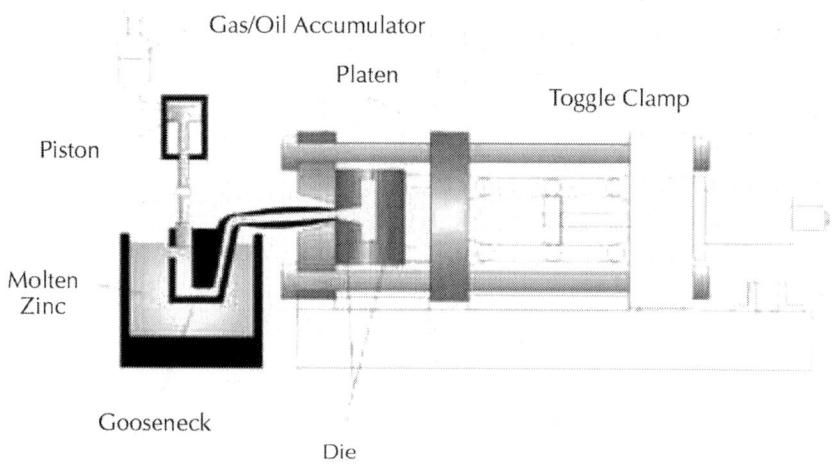

Figure 15: Schematics of hot chamber die casting machine with major components exposed to liquid magnesium alloys [84].

It is clear that more harsh conditions exist in the hot chamber machine. This is mainly because during all processing cycles, the hot chamber has a gooseneck completely submerged in molten magnesium. As a result, reduction in material strength caused by high temperature occurs, which has to be compensated by an increased size (Fig. 16). As such, the goosenecks are relatively robust. The massive component can be made of a casting using steel with high heat resistance, e.g. 1.2888 types. Due to issues with casting defects, it is frequently being replaced with wrought martensitic stainless steels, e.g. of AISI 420 type. Forging is an attractive alternative for the cast goosenecks. In the forging process, a solid billet can be shaped on open die presses, which consolidate the ingot center and eliminate porosity. It delivers a finished product that is more reliable and lasts longer [53]. Although the chemistry of AISI 420 steel provides lower resistance to tempering than DIN 1.2888, its high integrity in a wrought state may contribute to overall better performance. Another component exposed to liquid magnesium is the machine nozzle (Fig. 17a). Liquid magnesium along with melt pressure and non-uniform temperature distribution (Fig. 17b) often contribute to premature nozzle failure as seen in Fig. 17c.

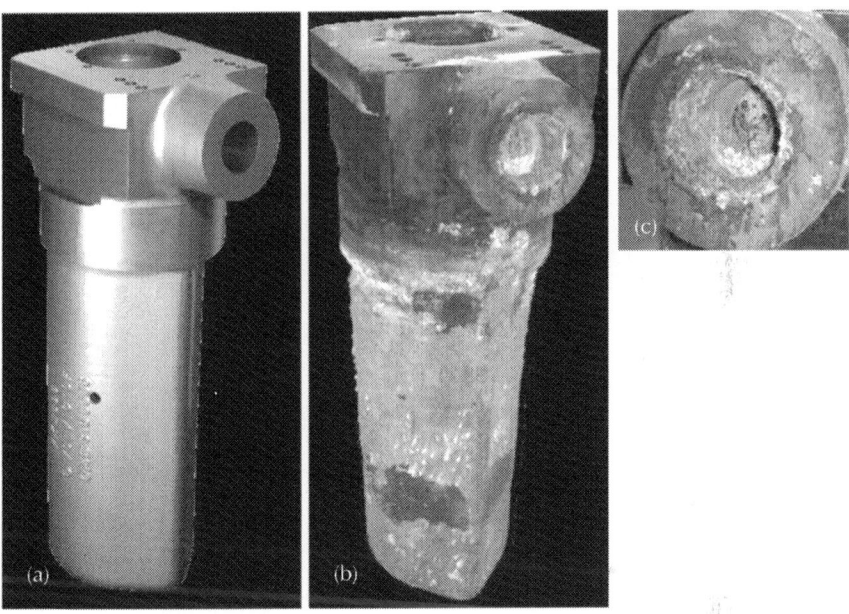

Figure 16: Goosenecks of hot chamber die casting machine: (a) part before exposure to liquid magnesium; (b, c) gooseneck after service with extensive corrosive attack [85].

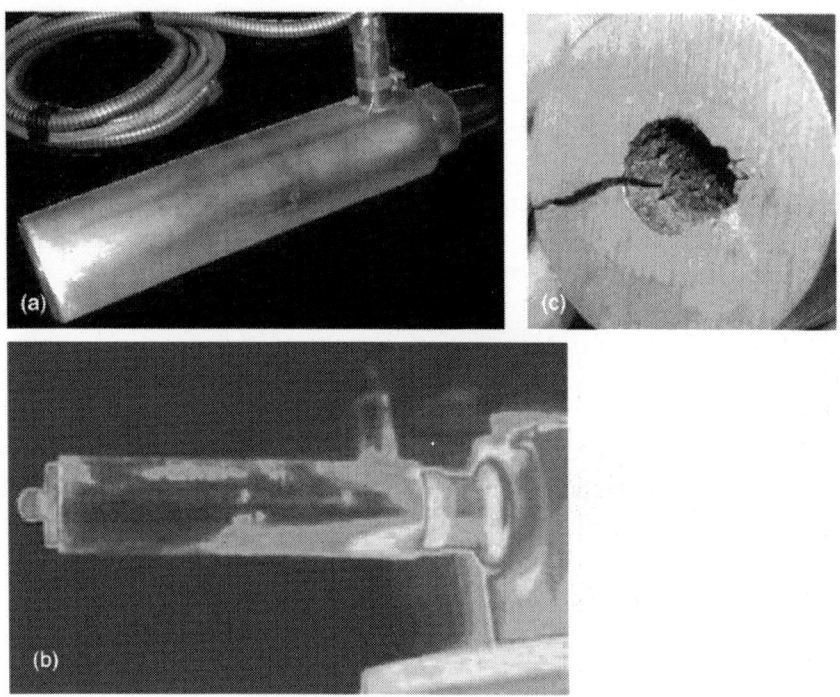

Figure 17: Nozzle of hot chamber die casting machine: (a) general view; (b) thermal profile showing temperature distribution; (c) section showing extensive corrosive attack and radial cracking [85].

Equipment of Novel Techniques of Magnesium Processing

In addition to high pressure die casting, there are a number of novel technologies at different advancement stages of commercialization, aimed at manufacturing net shape components from liquid or semisolid precursors [54] [55] [56] [57] [58] [59]. Of particular interest is equipment for semisolid processing of magnesium alloys and their composites. The key feature of machines for semisolid processing operating in thixo-mode, i. e., exploring only partial melting before injection, is reduced service temperature. Depending on the solid fraction targeted, the temperature reduction may be substantial. Instead of the overheated liquid used in die casting with a temperature of 670

°C, the process may require only 580 – 600 °C. This reduction may be of critical importance, affecting performance or even the applicability limit of some materials. At the same time, there is an essential difference in operating mode between die-casting and injection molding [1]. As opposed to short time intermittent contact, encountered between the shot sleeve of a cold chamber die casting machine and a molten metal, during injection molding the exposure of some machine components to molten alloy is permanent, reaching continuously in one interval up to thousands of hours. Moreover, while the shot sleeve of a die-casting machine has cooling channels, an injection molding barrel is extensively heated from outside. As a result, a substantial difference in service conditions renders the majority of solutions from melting and casting equipment not applicable to novel techniques of magnesium processing. Some materials with a sufficient combination of strength and corrosion resistance are still being researched [60].

MANUFACTURING TECHNIQUES UTILIZING MATERIAL COMBINATIONS

The majority of materials maintaining strength at high temperatures are not chemically resistant to liquid magnesium. In fact, alloys with the best ability to maintain strength at high temperatures, easily react with molten magnesium. Therefore, separating thermal and corrosive factors increases the choices of material selection during the design of the magnesium processing equipment. The most typical techniques applicable to explore material combinations are characterized below.

Mechanical Shielding

An interference-fit type connection may be used with components of specific shapes to cover the surface of an alloy which is prone to chemical attack by liquid magnesium while another alloy may be resistant against such attack. The technique applicable is called shrink-fitting and explores thermal expansion and contraction of metals to create a strong joint of a mechanical nature between two pieces, where one of each is inserted into the other. While in the past blow torches, hot plates and oil baths dominated as a heat source, the electric resistance and electromagnetic induction techniques are the commercially viable ones at present. Modern equipment allows uniform distribution of temperature with no danger of overheating by continuous monitoring the component temperature and adjusting its own heating power.

The typical example involves shrink-fitting of two tubes where the larger one is preheated to expand its diameter. At the same time, the inner tube remains at room temperature or is additionally cooled using, e.g., dry ice (CO_2 at a temperature of $-180\ ^\circ C$) to further reduce its diameter (Fig. 18). After an insertion of one tube into another and cooling to room temperature, both tubes are joined together. The interference value within the connection should be calculated to avoid exceeding the yield stress in both materials [61]. After excessive heating during service, simultaneous expansion of both tubes may eliminate interference and release preload. Thus, the high temperature applicability of such a connection depends on differences in coefficients of thermal expansion. The outer material cannot expand substantially more than the inner one. The connection is penetrable to air; so high temperature oxidation should be taken into account.

External sleeve: high temperature strength

Insert: liquid metal corrosion, wear, erosion

Figure 18: Concept showing tubes joined by a shrink fit method.

An application of shrink fitting is suggested for the manufacture of barrels for injection molding of magnesium [60]. In this design, an external shell with high temperature strength made of Inconel 718 superalloy is connected through shrink fitting with an internal liner made of Stellite 12 alloy. The latter one is corrosion resistant to liquid magnesium but it does not have sufficient ductility and tensile strength to withstand internal melt pressure.

Plasma Transferred Arc (PTA) Cladding

Cladding is a deposition of an alloy on the metallic substrate in order to modify the structure and physical characteristics of its surface required for certain applications. The most commonly used cladding techniques utilize the plasma transferred arc (PTA) and laser. PTA cladding is performed by melting of metallic, ceramic or cermets powder in a plasma arc before deposition. Typical characteristics include high deposition rate and minimum penetration into the base (<5%). Layers up to 5 mm in a single pass can be formed with very low losses of the deposited powder. PTA gradually replaces older methods of conventional weld overlays with tungsten inert gas (TIG) or metal inert gas (MIG). The generally simple concept may become complex when applied to specific material combinations. Of particular importance for processing liquid magnesium is cladding alloys having high temperature strength with alloys resistant to wear and corrosive attack by molten magnesium.

Weld Cladding of Ni-Based with Co-Based Superalloys

Inconel 718 has good weldability and preferred techniques include gas tungsten arc, plasma arc and electron beam welding [62]. The process is performed on annealed material but the age-hardened alloys are also weldable. The post-welding treatment includes most often annealing and aging but in many instances aging is sufficient if material was welded in annealed condition. Preheating is not usually required, excluding warming up to prevent moisture condensation.

The typical defects during welding involving Inconel 718 alloy include:

- Solidification cracking, which occurs within a newly formed weld when the semisolid region experiences tensile stress and the high fraction of solid restricts the flow of liquid to fill interdenfritic regions;

- Grain boundary liquation cracking or heat affected zone fissuring, which occurs within the heat affected zone as a result of local dissolution of grain boundary phases. Under rapid heating, the grain boundary phases area that is unable to dissolve fully into the surrounding matrix leads to formation of a low melting point eutectics and a melting of the grain boundary region;

- Strain age cracking, which occurs during a post-weld heat treatment or high temperature service as a result of residual or applied stress. The microstructural image usually shows intergranular microcracks in either weld or heat affected zones due to precipitation and hardening of the alloy and transfer of stress on grain boundaries where hard precipitates may act as a crack nucleolus.

The Co-based alloys with high carbon content are difficult to weld. To prevent cracking, the alloy should be preheated and maintained at 540 °C minimum. The cooling rate should also be slower. While cladding Ni-base with Co-based alloys the requirements in regards of heat control are contradictory: fast cooling rate for Ni-based alloy to avoid hot cracking versus slow cooling rate for Co-based alloy to avoid cold cracks. As a possible solution it is suggested to use an underlayer. For this purpose a pure Ni or Inconel 82 layer with a thickness of 1-2 mm is recommended [63].

An application of PTA cladding for equipment used with liquid magnesium is specified in Ref. [64]. It was proposed that barrel head be manufactured of a Ni-based superalloy, such as Inconel 718, the melt channel clad with low C content, Co based compact (crack free) Stellite 21. The seal surfaces of the barrel head were clad with high C content Co based superalloy such as Stellite 12. The microstructure of Inconel 718 cladded with Stellite 21 is shown in Fig. 19. With Stellites in general, the primary face cubic centered cobalt dendrites are surrounded by a network of eutectic lamellae composed of cobalt and eutectic M_7C_3 carbides. The composition of M_7C_3 depends on alloy type and cooling conditions with a typical example of $Cr_{0.85} Co_{0.14} W_{0.01})_7C_3$ [65]. In addition to pure alloys it is also formed in a mixed zone, where chemistry of molten alloys combines.

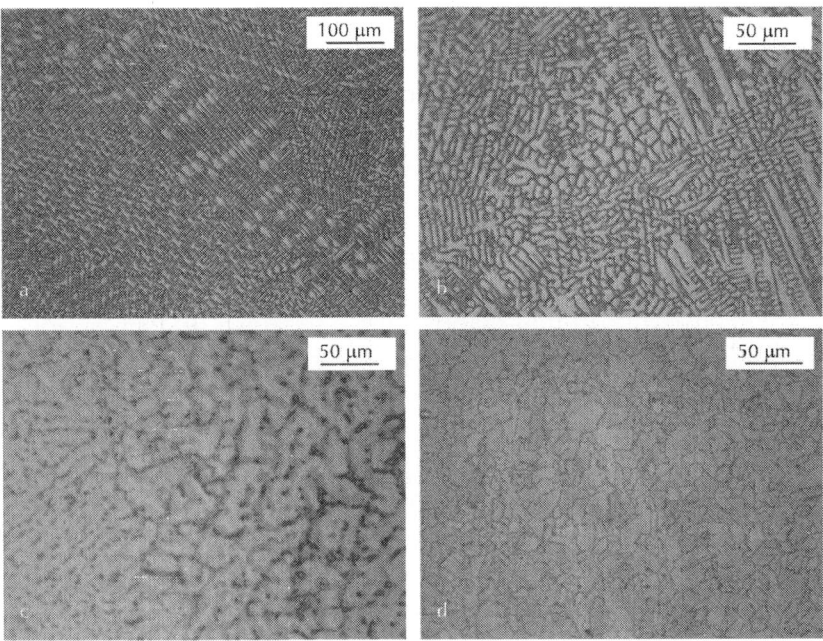

Figure 19: PTA cladding of Inconel 718 with Stellite 12 alloy: (a, b) microstructure of surface clad of Stellite 12; (c) transient zone between base and clad; (d) Inconel 718 in the substrate.

Laser Cladding

Laser cladding represents the fusion of a different material to a substrate surface while ensuring the metallurgical bond with minimal melting of the substrate and chemistry dilution, as well as the small heat affected zone. It can be conducted in a single or two-stage process. In a single stage, called the blown powder cladding, the alloy powder, transported into the interaction zone between the laser beam and substrate, is subjected to heating. Melting starts at the substrate surface and powder particles form a pool. Selection of the laser energy allows to control the substrate melting. In the generally simpler, two stage cladding; the powder is first pre-deposited on the substrate. To keep the powder on the surface, various binders are applied. Then, the powder is scanned by the laser beam while covered by an inert gas. Three distinct stages during melting of the powder are distinguished [66]:

- First, the powder is rapidly melted by the laser before the melt gets in contact with the substrate;
- When the melt reaches the substrate surface, it solidifies due to a rapid flow of the heat into the substrate by conduction. There is no movement of the melt-liquid interface into the substrate;
- Further applying the laser energy will move the former melt-solid interface deeper into the substrate.

Benefits of laser cladding include [67]: (i) very small heat-affected zone, which results in tiny deformation and stress; (ii) can be applied to virtually all metal materials; (iii) operation time is significantly shortened compared to arc welding; (iv) final product has high dimensional accuracy and integrity.

Laser cladding has application potentials in magnesium casting, e.g. in extending life time of dies/molds by increasing their wear resistance with hard surface layers to reduce erosion. By combining high wear with high tensile strength and high ductility, thermal or stress induced cracking during the casting process can be reduced. Since commonly used hot working tool steels have limited wear resistance, laser cladding is a very useful technique to improve their surface properties by multi-graded layers [68]. An application of laser cladding to depositStellite 1 on AISI 4340 steel is described in Ref. [69]. Cobalt-based/carbide type alloys are well-known for their "hot hardness" and are extensively used for hardfacing of components made of conventional steels. The main challenge in deposition of Stellite 1 using laser cladding is crack sensitivity of this alloy during hardfacing process. To reduce the possibility of cracking, preheating of the substrate prior to the deposition process was tested to be effective. Investigation of die soldering during high pressure die casting led to a promising treatment with the laser clad Fe-W alloys [39]. An increased W content in the powder mix led to reduction in formation of intermetallic phases on die surface due to negligible possibility of a reaction between W and Al from Mg alloy during casting.

Centrifugal Cast Cladding

During centrifugal casting a liquid metal is poured into a rotating tube-like crucible and after solidification removed from it [70]. By replacing the crucible with the sleeve material, the process can be used for cladding of its inner surfaces. In addition, the liner material may be pre-applied as powder and subsequently melted while inside the sleeve [71]. The basic requirement is a safe difference between melting ranges of both alloys. The downside of this technique is that the cladded part has to be preheated to high temperatures causing coarsening of the microstructure and property reduction. An example of cladding the Inconel 718 with Stellite 12 is shown in Fig. 20. The metallurgical bonding formed assures strong connection. Due to severe overheat the grain growth occurred from an initial number 7 to number 00 according to the ASTM scale. Subsequent aging does not produce optimum phase morphology. On the other hand, a requirement to keep the temperature as low as possible causes incomplete melting of Stellite and individual powder particulates still remain within the liner.

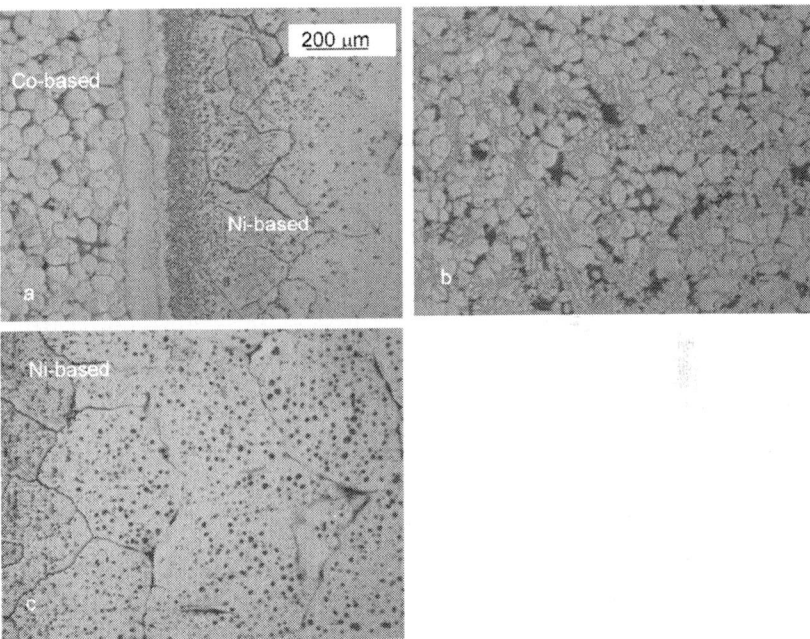

Figure 20: Microstructure of centrifugal inlays casting of Co-based alloy inside Ni-based shell: (a) interface region; (b) Co-based inner layer; (c) outer layer of Ni-based alloy.

HIP-Cladding

Hot isostatic pressing (HIP) is a densification process for both encapsulated powders and pre-formed parts, e.g. castings. It was developed in 1955 at Battelle Laboratories, Columbus, Ohio to bond components of small Zircaloy-clad pin-type nuclear fuel elements while maintaining strict dimensional control. The technique involves the simultaneous application of a high pressure and elevated temperature. The isostatic nature of pressure is achieved due to its application through a gas, most frequently inert, and it should be distinguished from a conventional unidirectional pressing. The isostatic pressure in HIP-ing arises from molecules or atoms of gas colliding with the surface of the object [72]. Under conditions of heat and pressure the encapsulated powder or sintered components are densified to improve properties.

HIP-ing creates homogeneous material with a uniformly fine grain size and near 100% density. Internal voids are healed and a strong metallurgical bond is created within the entire volume. There also exists the Liquid HIP-ing, typically used for densification of castings, where to lower cost, molten salts and mechanically generated pressures are explored. HIP should also be distinguished from cold isostatic pressing (CIP) which is a compaction of powders enclosed in an elastometer mold. During manufacturing of near net shape components from a single alloy, a powder mixture of several elements is placed in a steel can container. Then air and moisture is removed from the powder by applying high temperature and vacuum. Finally, the container is sealed and HIPed. During application of HIP-ing for manufacturing of bimetallic structures, two different chemistry powders are placed into two separated chambers of the steel container. After HIP-ing is completed and external shape machined, the both alloys are separated by the interface layer, being the container wall. There is a metallurgical bond between the steel layer and alloys on its both sides. HIP ing is used to manufacture alloys such as Ni-based and Co-based superalloys [73], Nb-30Ti-20W and also near-net shape components used for processing of molten magnesium. Besides, HIP-ing can be used for cladding selected surfaces of wrought substrates with another alloy [74]. In this process, a portion of the container is the wrought alloy. After HIP-ing, the thin steel container is machined out, leaving a bimetallic structure of wrought sleeve and powder cladding. In this technique, the HIP-ing behavior in the formation of an interfacial diffusion bonding among dissimilar materials is explored. A drawback of this type of cladding is a deteriorating effect of high temperatures required for HIP-ing on the structure of wrought alloy. In principle, the process may employ the same alloys as described above for weld cladding.

SURFACE MODIFICATIONS

The purpose of surface engineering is to enhance properties of superficial layers of materials by changing their chemistry and/or structure. There are two essentially different cases of surface protection, which depend on the substrate (base) material and its compatibility with molten magnesium. For materials resistant to chemical attack by molten magnesium, surface engineering aims at improving this resistance. In this case the coating failure does not lead to the catastrophic failure of the part. For materials not resistant to molten magnesium, surface engineering offers basic protection to the substrate. In this case, however, failure of the coating leads to catastrophic failure of the entire part.

Thermochemical Diffusion Treatments

Thermochemical diffusion is a surface treatment where the chemistry of superficial layers of materials is altered by introduction of some chemical elements from outside [75]. The atoms introduced combine with alloy elements, thus modifying existing phases and forming new ones. The major processes applicable to protection against liquid magnesium include nitriding, nitrocarburizing and boriding.

Nitriding

Nitriding is a process of enriching the surface layer with nitrogen. In case of steel substrate a compound layer is developed on the surface constituted primarily of iron nitrides, Fe_4N called γ' (gamma prime) or $Fe_{2-3}N$, called ε (epsilon) The structure of this compound zone is mainly determined by the underlying diffusion zone, which serves as a transition zone with declining hardness from the high hardness at the surface to that of the core of the material. There are several distinct approaches to nitriding [75]:

- Conventional gas nitriding, carried out in partially dissociated ammonia gas at 500-600 °C. The disadvantage is a lack of adequate control of the nitrogen concentration;

- Gas nitriding where nitriding potential is continuously controlled taking into account varying compositions of the nitriding atmosphere during each stage of the cycle (Nitreg);

- Ion nitriding where a plasma glow process ionizes nitrogen gas with heat causing the positive ions of nitrogen created to be drawn onto the surface of the component forming a uniform layer.

Ferritic Nitrocarburizing

Nitrocarburizing is considered as a complementary process to nitriding and can be carried out in liquid, gaseous and plasma environments. During nitrocarburizing, the steel surface is enriched simultaneously with nitrogen and carbon. The process is carried out at 550 – 580 °C and depending on the composition of the base material and exposure time the penetration depth reaches from 200 μm to 1000 μm with a surface hardness from 700 to 850 HV. Assuming sufficiently high activities of carbon and nitrogen, the compound layer, formed at the surface, consists predominantly of ε and/or γ' phases. Beneath it, there is a diffusion zone with N and C atoms dissolved interstitially in the ferrite lattice. While the compound layer brings a combination of resistance to wear and atmospheric corrosion, the diffusion zone improves the endurance limit [76].

It should be stated that steel nitriding is widely used for cold chamber die casting shot sleeves, die/mold surface, die pins, piston rings etc. Its effectiveness is limited, at least for some applications, first because chemistry of steels used for machinery of magnesium processing are not optimized for nitriding or nitrocarburizing. Moreover, high service temperature may diffuse away nitrogen from the surface layer, deteriorating its properties over time. An example of an H13 commercial piston ring from a cold chamber die casting machine is shown in Fig. 21. Although due to temperature experienced in service the initial case hardness, along with hardness of the base substrate is reduced, the overall ring performance after nitriding is substantially improved.

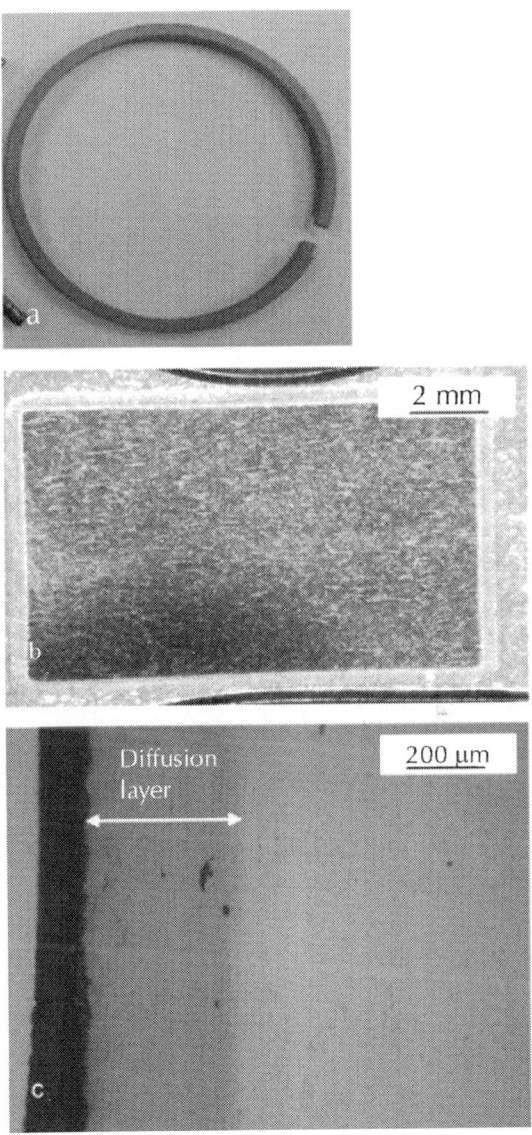

Figure 21: Nitrided layer on the H13 steel used for die cast piston ring: (a) piston ring; (b) macro cross section showing thick nitrided layer; (c) micro image showing the nitrogen penetration range.

Boriding (Boronizing)

Boriding means an enrichment of the surface layer of the material with the elemental boron. Unlike nitriding, boriding is carried out at significantly higher temperatures, typically between 800 and 1000°C [77]. Various implementation technologies include packing with solid mixtures of boron carbide and borax as the boron source, molten salt performed using anhydroux borax mixed with reducing agents, electrolytic carried out in an argon atmosphere using borax-based melts, plasma and vacuum boriding [78]. The surface hardness is in the range of 1500-2000 HV, which exceeds nitriding and nitrocarburizing. The technique may be applied to ferrous and nonferrous metals and alloys, e.g. Ti, Ni or Co. Depending on the substrate chemistry, the surface layer forms different compounds. For steel, Fe_2B is the major compound formed. For Ni the major compound is Ni_3B while boriding Co produced both Co_4B and Co_3B. The boriding treatment was explored to prevent soldering during Mg die casting.

Thermo-Reactive Diffusion (TRD)

The vanadium carbide thermo reactive diffusion creates a 5 – 10 μm thick surface layer with a hardness of 3200 – 3800 HV. The process is conducted in a packed or molten salt environment, where active vanadium enters the surface and combines with carbon atoms from steel to form vanadium carbides [79]. Vanadizing is performed at temperatures of 800 – 1000 °C. Thickness is well controllable by adjusting diffusion time and temperature for a given environment and steel chemistry. The layer has a very strong metallurgical-diffusion bond to the substrate, providing high peeling resistance. Thermo-reactive diffusion was originally developed to improve wear resistance. It has a thermal expansion mismatch with chromium hot work tool steel AISI H13. According to some data minimum carbon content should be 0.3%. The process is effective in die casting operations and explored especially in Europe and Japan. The major parts treated include die casting cores and pins. According to [80], for Al casting, TRD coatings led to higher increase in tool life than nitriding. As a partial explanation is given that VC resists aluminum diffusion through the surface layer.

Coatings

According to general selection rules the coating should have high hardness and strength and moderate ductility. Its thermal conductivity, melting point and density should also be high while the coefficient of thermal expansion and coefficient of friction should be low [81]. Due to harsh environment of molten Mg, of a large number of coating techniques and coating chemistries, only several were found to provide improvement. Below is provided their brief characterization.

Physical Vapor Deposition (PVD)

PVD includes coating techniques where the transport of atoms or molecules to the coated surface is accomplished by a physical process. The technique of line-of-sight coating of complex geometries is difficult. The low deposition temperature of 100 – 500 °C minimally affects the substrate structure. Due to low temperature, however, it is difficult to achieve great adhesion. The coating chemistries applicable to Mg processing are given in Table 4 with indications of their thermal stability.

Table 4: Selection of PVD coatings applicable for components of magnesium processing equipment exposed to liquid magnesium alloys [89] [90] [91]

Name	Formula	Colour	Thermal Stability,°C	Hardness, HV	Typical thickness, μm	Characteristic features
Titanium nitride	TiN	Gold	600	2300	2-5	Used for moderate abrasions, ductile hard coat
Titanium carbonitride	TiCN	Grey-pink	410	3000	2-5	Shock resistance
Titanium aluminum nitride	TiAlN	Brown	800	3500	2-5	Extreme heat resistance
Zirconium nitride	ZrN	Yellow-gold	650	2600	2-5	Excellent lubricity
Chromium nitride	CrN	Silver	1750	695	2-10	Good wear resistance

Titanium diboride	TiB2	Silver	900	4000	1-2	High corrosion resistance, chemical stability at elevated temperatures
Titanium boron nitride	TiBN	Silver	800	3500	1-5	Heat-checking resistance, abrasive wear resistance

The general finding in regards to selected components of Mg processing equipment is that coatings provide substantial improvement against sticking of Mg to the surface and reduction of an overall degradation of a corrosive nature. This often allows the elimination of an acidizing step during cleaning and caused by this hydrogen ingress into the base. At the same time, however, 2-5 micron thick TiAlN coatings with mechanical bond to the substrate show severe limitations since they:

- Do not provide long lasting improvement in the extent of wear between metallic couples;
- Do not provide a lasting chemical barrier with high integrity to separate the base material from liquid magnesium.

The barrier integrity is lost not only by the above named surface degradation mechanisms, but also by localized wear caused by hard particles which may enter the melt stream. During die casting of magnesium, tests of several PVD coatings [39], TiN on ground H13 steel were showing good performance. In contrast, PVD coating of CrN was dissolving with intermetallics formed on its top, especially in Mg alloys with higher Al content.

Chemical Vapor Deposition (CVD)

The CVD process uses chemical reactions to deposit coatings on the substrate. In general, the process is carried out at higher temperatures than PVD, up to 2200 °C which provides excellent adhesion. The temperature requirement reduces markedly a number of applicable substrates. Since the nature of CVD is non-line-of-sight, it can be used to evenly coat complex geometries and internal surfaces. The typical chemistries deposited with CVD include CrC, TiN and VC. Applicability to liquid Mg of CVD coating itself is the same as discussed above for PVD method.

Thermal Spray

During thermal spraying, the wire or powder materials are melted into droplets, and then propelled onto the selected substrate. Upon impact, they form platelets that bond to the surface, creating a dense coating with no alteration to the substrate structure. The technique allows the deposition of pure metals, alloys, intermetallics, carbides and ceramics. There are several techniques applicable for deposition spray coatings:

- Combustion flame spray, achieved by burning a mixture of oxygen and fuel gas in a torch having a flame-accelerating nozzle. Powder is injected into the nozzle by the carrier gas, where it melts and is projected to the surface;

- Electric arc spray, achieved by energizing two wires of the coating material at different electrical potentials. Molten particles are generated by arcing the wire tips which are then atomized and accelerated towards the substrate by a compressed gas;

- Hypervelocity oxygen fuel spray (HVOF) is achieved by burning a pressurized mixture of the fuel gas and oxygen. Powder is fed into the stream of hot gases and discharged through a flow expansion zone;

- Plasma spray, achieved by exploring the heat transfer for electric arc to a plasma-forming gas. In the spray device the gas flow contains an axial stick cathode while the nozzle forms the anode. Heated to high temperatures gas ionizes to plasma. Powder, injected into the exit melts and is accelerated by hot gases towards the surface. Special cooling techniques keep the surface temperature low.

An example of thermal spray coating is shown in Fig. 22a. The top end represents a 0.13 mm thick ceramic coating of YSZ (yttria stabilized zirconia) which exhibits very low thermal conductivity and provides thermal insulation. The lower cylindrical portion is covered with a 0.2 mm thick metallic coating of NiCr6Al providing oxidation resistance up to 980 °C. In some cases the ceramic coatings are too brittle to withstand mechanical stress. To minimize their brittle nature the ceramic coatings may be used in a combination with metallic support. An example of the solution is given in Fig. 22b. The metallic grid provides support and reduces surface contact while the ceramic coating, filling gaps between them, reduces the heat transfer.

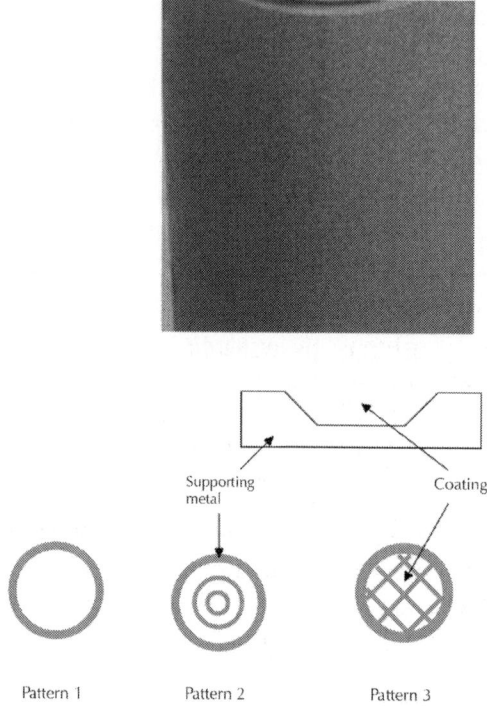

Figure 22: Ceramic coating of YSZ (yttria stabilized zirconia) deposited by thermal spray (a) macro view of the coating; (b) concept of the surface grooving pattern to support the brittle ceramic coating during impact.

Hybrids of Surface Treatments

Due to harsh service conditions of Mg processing, in some cases a single surface treatment may not be sufficient. It is claimed that single coating cannot eliminate completely the deteriorating effect of liquid magnesium alloys. Thus, in order to maximize the substrate protection, two or more surface modification techniques are combined. The multilayer hybrids combine advantages of individual layers. A concept, explaining the mechanism of substrate protection by multilayer coatings is shown in Fig. 23.

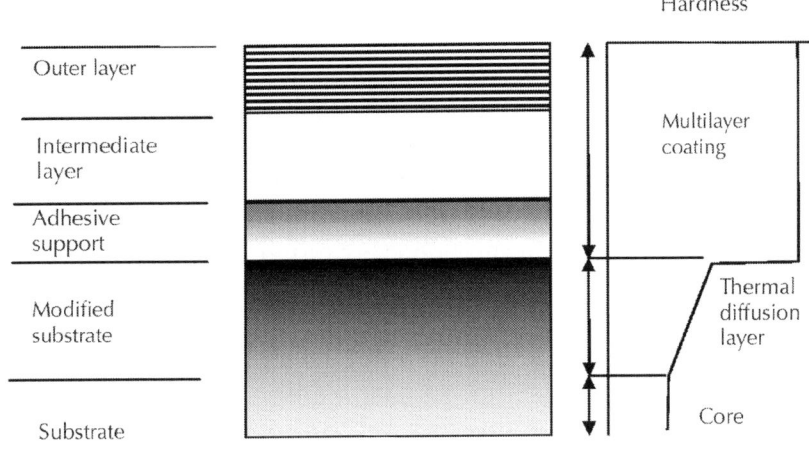

Figure 23: Concept of multilayer coatings deposited by PVD onto a steel surface enhanced by thermochemical diffusion.

A Combination of Thermal Diffusion Layers with Coating

The performance of coatings, especially those thin ones, depends on the properties of the substrate. It is clear that a hard coating will not last on soft substrate since it will crack and spall. In order to improve coating performance, substrates are subjected to special treatments. The most common is thermochemical diffusion treatment e.g. nitriding. Then, a coating is deposited on the top of the diffusion layer. An additional benefit is that in case of local failure of the coating, a diffusion layer may still protect the substrate. While this combination generally provides improvement, for some cases the substrate effect is negative. According to [39], TiN and CrN PVD coatings deposited on nitrided H13 steel showed poor adhesion and during Mg die casting process peeled off from the substrate. The poor performance was attributed to the decomposition of Fe-N nitride layer at increased temperatures due to diffusion of nitrogen.

Multilayer Coatings

The progress in deposition technology allows the automatic control of the process so multilayer coatings may be formed. In practice, individual sub-layers will differ in terms of chemistry and structure. The properties of such a conglomerate exceed the properties of individual sub-layers. This method represents new type of surface treatment. In addition to coating conglomerates, multi-layer coating can be combined with diffusion treatment of the substrate. During trials with die casting of magnesium [39], the best performance was reached with multilayer PVD TiN/CrN coatings which completely suppressed formation of intermetallics. In contrast, duplex TiN/CrN coatings did not show an improvement over single layer of TiN. When deposited on nitrided steel surface, decomposition of the Fe-N white layer due to internal diffusion reduced the coating performance.

SUMMARY

The high temperature and corrosive attack of molten magnesium alloys impose a challenge on the selection of materials applicable for the hardware used for their processing. This challenge is magnified in the case of novel processing techniques where performance of materials used in conventional equipment of magnesium casting and handling is not satisfactory. Although modern metallurgy provides solutions to present requirements, the research continues to develop alloys capable of increasing the processing temperatures and extending the service life time of processing equipment.

REFERENCES

1.   F. Czerwinski, Magnesium Injection Molding, New York: Springer Verlag, 2008.

2.   F. Czerwinski, "Surface oxidation of magnesium alloys during liquid state processing," Die Casting Engineer, vol. 57, no. 5, pp. 16-19, 2013.

3.   F. Czerwinski (Ed.), Magnesium Alloys, Corrosion and Surface Treatment, Rijeka: INTECH, 2011.

4. F. Czerwinski, "Oxidation characteristics of magnesium alloys," The Journal of The Minerals, Metals and Materials Society, vol. 54, no. 12, pp. 1477-1483, 2012.

5. F. Czerwinski, "Controlling the ignition and flammability of magnesium for aerospace applications," Corrosion Science, vol. 86, pp. 1-16, 2014.

6. F. Czerwinski, "Overcomming barriers of magnesium ignition and flammability," Advanced Materials and Processes, vol. 172, no. 5, pp. 28-31, 2014.

7. L. Liu (Ed), Welding and Joining of Magnesium Alloys, Woodhead Publishing, 2010

8. F. Czerwinski (Ed.), Magnesium Alloys-Design, Processing and Properties, Rijeka: INTECH, 2011.

9. D. Bradwell, H. Kim, A. Sirk and D. Sadoway, "Magnesium-antimony liquid metal battery for stationary energy storage," Journal of the American Chemical Society, vol. 134, no. 4, pp. 1895-1897, 2012.

10. K. Papis, J. Loeffler and P. Uggowitzer, "Interface formation between liquid and solid Mg alloy-An approach to continuously metallurgic joining of magnesium parts," Materials Science and Engineering A, vol. 527, pp. 2274-2279, 2010.

11. J. Zhang, "A review of steel corrosion by liquid lead and lead-bismuth," Corrosion Science, vol. 51, pp. 1207-1227, 2009.

12. X. Liu, E. Barbero, J. Xu, M. Burris, K. Chang and V. Sikka, "Liquid metal corrosion of 316L, Fe3Al and FeCrSi in molten Zn-Al baths,"Metallurgical and Materials Transactions A, vol. 36, pp. 2049-2058, 2003.

13. T. Wetzel et al, "Liquid metal technology for concentrated solar power systems: contribution by the German research program," AIMS Energy, vol. 2, no. 1, pp. 89-98, 2014.

14. W. Manly, Fundamentals of liquid metal corrosion, Oak Ridge: Oak Ridge National Laboratory, 1956.

15. S. Shin et al, "A study on corrosion behavior of austenitic stainless steel in liquid metals at high temperature," Journal of Nuclear Materials, vol. 422, pp. 92-102, 2012.

16. J. Zhang, P. Hosemann and S. Maloy, "Models of liquid metal corrosion," Journal of Nuclear Materials, vol. 404, pp. 82-96, 2010.

17. T. Auger, Z. Hamouche, L. Medina-Almazan and D. Gorse, "Liquid metal embrittlement of T91 and 316L steels by heavy liquid metals: A fracture mechanics assessment," Journal of Nuclear Materials, vol. 377, pp. 253-260, 2008.

18. K. Meshinchi Asl and J. Luo, "Impurity effect on the intergranular liquid bismuth penetration in polycrystalline nickel," Acta Materialia, vol. 60, pp. 149-165, 2012.

19. S. Keller and A. Gordon, "Experimental study of liquid metal embrittlement for the aluminum 7075-mercury couple," Engineering Fracture Mechanics, vol. 84, pp. 146-160, 2012.

20. P. Lejcek, Grain Boundary Segregation in Metals, Berlin, Heidelberg: Springer Verlag, p. 183, 2010.

21. R. Clegg and R. Jones, "Liquid metal embrittlement in failure analysis," Soviet Materials Science, vol. 27, no. 5, pp. 453-459, 1991.

22. K. Ina and H. Koizumi, "Penetration of liquid metals into solid metals and liquid metal embrittlement," Materials Science and Engineering A, Vols. 387-389, pp. 390-394, 2004.

23. J. Luo, H. Cheng, K. Meshinchi ASL, C. Kiely and M. Harmer, "The role of a bilayer interfacial phase on liquid metal embrittlement," Science, vol. 333, no. 6050, pp. 1730-1733, 2011.

24. A. Legris, J. Vogt, A. Verleene and I. Serre, "Wetting and mechanical properties, a case study: liquid metal embrittlement of a martensitic steel by liquid lead and other liquid metals," Journal of Materials Science, vol. 40, pp. 2459-2463, 2005.

25. H. Migge, "Thermodynamic stability of ceramic materials in liquid metals illustrated by beryllium compound in liquid lithium," Journal of Nuclear Materials, vol. 103, no. 1-3, pp. 687-692, 1982.

26. W. Cook, Corrosion resistance of various ceramics and cerments to liquid metals, Oak Ridge: Oak Ridge National Laboratory, 1960.

27. R. Sangiorgi, "Corrosion of ceramics by liquid metals," Corrosion of Advanced Ceramics, NATO Science Series E, vol. 267, pp. 261-284, 1994.

28. E. Shchukin and V. Bravinskii, "Reduction in strength of ceramic materials under the influence of molten metals," Soviet Materials Science, vol. 4, no. 3, pp. 216-219, 1968.

29. A. Nayeb-Hashemi, J. Clark and L. Swartzendruber, "The Fe-Mg (Iron-Magnesium) system," Bulletin of Alloy Phase Diagram, vol. 6, no. 3, p. 235, 1985.

30. P. Merica and R. Waltenberg, Technical Papers of, National Bureau of Standards (USA), vol. 19, pp. 155-182, 1925.

31. A. Nayeb-Hashemi and J. Clark, "The Mg-Ni (Magnesium-Nickel) system," Bulletin of Alloy Phase Diagram, vol. 6, no. 3, pp. 238-239, 1985.

32. K. Micke and H. Ipser, Monatsh. Chemistry, vol. 127, pp. 7-13, 1996.

33. P. Bagnoud and P. Peschotte, Z. Metallkunde, vol. 127, pp. 114-120, 1978.

34. Q. Hong and F. d'Heurle, "The dominant difusing species and initial phase formation in Al-Cu, Mg-Cu and Mg-Ni systems," Journal of Applied Physics, vol. 72, p. 4036, 1992.

35. C. Tsao and S. Chen, "Interfacial reactions in the liquid diffusion couples of Mg/Ni, Al/Ni and Al/ (Ni)-Al2O3 systems," Journal of Materials Science, vol. 30, pp. 5215-5222, 1995.

36. C. Roberts, Magnesium and Its Alloys, New York: John Wiley and Sons, 1960.

37. M. Yan and Z. Fan, "Durability of materials in molten aluminum alloys," Journal of Materials Science, vol. 36, pp. 285-295, 2001.

38. J. Viala, D. Pierre, F. Bosselet, M. Peronnet and J. Bouix, "Chemical interaction processes at the interface between mild steel and liquid magnesium of technical grade," Scripta Materialia, vol. 40, no. 10, pp. 1185-1190, 1999.

39. C. Tang, Soldering in magnesium high pressure die casting and its prevention by surface engineering, Hawthorn, VIC, Australia: Swinburne University of Technology, 2007.

40. C. Brooks, Heat Treatment, Structure and Properties of Non-ferrous Alloys, Metals Park, OH: ASM International, 1982.

41. E. Agnion and B. Bronfin, in Proceedings of the 3-rd International Magnessium Conference, p. 313, The Institute of Materials, London, 1997.

42. R. Bruscato, "Liquid metal embrittlement of austenitic stainless steel when welded to galvanized steel," Welding Research Supplement, vol. December, pp. 455-s-460-s, 1992.

43. H. Martinz, B. Nigg and A. Hoffmann, "The corrosion behavior of refractory metals against molten and evaporated zinc," in The 17th Plansee Seminar, Reutte, Austria, 2009, RM 56A1-7.

44. S. Wilson, A. Kvithyld, T. Engh and G. Tranell, "Oxidation of manganese-containing aluminum alloys studied by SEM," Materials Science Forum,Vols. 794-796, pp. 1095-1100, 2014.

45. C. Tang, M. Jahedi and Brandt, "Investigation of the soldering reaction in magnesium high pressure die casting dies," in 2002 International Tooling Conference, Karlstad, Sweden, 2002.

46. D. Pierre, J. Viala, M. Peronnet, F. Bosselet and J. Bouix, "Interface reactions between mild steel and liquid Mg-Mn alloys," Materials Science and Engineering A, vol. 349, pp. 256-264, 2003.

47. D. Pierre, M. Peronnet, F. Bosselet, J. Viala and J. Bouix, "Chemical interaction between mild steel and liquid Mg-Si alloys," Materials Science and Engineering B, vol. 94, pp. 186-195, 2002.

48. D. Pierre, F. Bosselet, M. Peronnet, J. Viala and J. Bouix, "Chemical reactivity of iron base substrates with liquid Mg-Zr alloys," Acta Materialia,vol. 49, pp. 653-662, 2001.

49. K. Humberstone, "Method of forming crucibles for molten magnesium". USA Patent 4,424,436, 3 January 1984.

50. K. Humberstone, "Crucibles for molten magnesium and method of forming". USA Patent 4,353,535, 12 October 1982.

51. J. Leland, "Method for protecting austenitic stainless steels from solvent attack by molten magnesium by forming crucible". US patent 5,227,120, 13 July 1993.

52. A. Ditze and C. Sharf, Recycling of Magnesium, Clausthal-Zellerfeld: Papierflleger Verlag, 2008.

53. "Forging improves magnesium die casting," Industry sourcing, [Online]. Available: http://www. industrysourcing. Com/ articles/255033. aspx. [Accessed 15 06 2014].

54. S. Moore, "Magnesium molding-technique expands options," Modern Plastics, vol. July, p. 33, 2002.

55. K. Kono, "Method and apparatus for manufacturing parts by fine die casting". US patent 5,983,976, 16 November 1999.

56. S. Moore, "Thixotropic molding broadens process capabilities," Modern Plastics, vol. March, pp. 24-30, 2002.

57. F. Czerwinski, "Near-liquidus injection molding process". US Patent 7,255,151, 2007 August 2007.

58. L. Rogal, F. Czerwinski, L. Litynska-Dobrzynska, P. Bobrowski, A. Wierzbica-Miernik and J. Dutkiewicz, "Effect of hot rolling and equal-channel angular pressing on generation of globular microstructure in semi-solid Mg-3%Zn alloy," Solid State Phenomena, Vols. 217-218, pp. 381-388, 2015.

59. F. Czerwinski, "Semisolid processing of magnesium alloys: microstructure-properties relationship," Soli State Phenomena, Vols. 217-218, pp. 3-7, 2015.

60. F. Czerwinski, "Metal molding conduit assembly of metal molding system". US Patent 20090107646, 30 April 2009.

61. T. Ozben, A. Yardimended and O. Cakir, "Stress analysis of shrink-fitted pin-pin hole connections via Finite Element Method," Journal of AMME,vol. 25, pp. 45-48, 2007.

62. J. Gordine, "Some problems of welding of Inconel 718 alloy," Welding and Research Supplement of AWS, vol. November, pp. 480s-484s, 1971.

63. J. Kelly, "Speciality alloys welding," Rolled Alloys, 2002. [Online]. Available: http://www. rolledalloys. ca. [Accessed 29 07 2014].

64. "Prior Art Database: IPCOM000012407D Bimetallic barrel head for processing corrosive metals and a method of manufacturing thereof," 03 05 2003 UTC USA. [Online]. Available: https:// priorart. ip. com. [Accessed 24 07 2014].

65. J. Cassina and I. Machado, "Low-stress sliding abrasion resistance of cobalt-based surfacing deposits welded with different processes," Welding Research Supplement AWS, vol. April, pp. 133s-138s, 1992.

66. H. Gedda, Laser cladding: an experimental and theoretical investigation, Ph. D. Thesis: Lulea University of Technology, 2004.

67. J. Wang, S. Prakash, Y. Joshi and F. Liou, "Laser aided part repair-a review," University of Missouri-Rolla, [Online]. Available: http:// utwired. engr. utexas. Edu/lff/symposium/proceedingsArchive/ pubs/Manuscripts/2002/2002-07-Wang. pdf. [Accessed 03 06 2014].

68. S. Ocylok, A. Weisheit and I. Kelbassa, "Functionally graded multi-layers by laser cladding for increased wear and corrosion resistance," Physics Procedia, vol. 5, pp. 359-367, 2010.

69. "Laser cladding of Stellite 1 on AISI 4340 steel," University of Waterloo, [Online]. Available: http://alfa. uwaterloo. Ca/ Research. Html. [Accessed 03 06 2014].

70. M. Gooover, Fundamentals of Modern Manufacturing: Materials, Processes, and Systems, Hoboken, NJ: John Wiley & Sons Ltd, 2010.

71. "Highly wear resistant barrels for extrusion and injection molding," Reiloy Reifenhausergroup, [Online]. Available: http:// www. reiloy. Com/fileadmin/reiloy/download/zylinder_A4_engl. pdf. [Accessed 25 07 2014].

72. H. Atkinson and S. Davis, "Fundamental aspects of hot isostatic pressing: an overview," Metallurgical and Materials Transactions A, vol. 31, p. 2981, 2000.

73. K. Pinnow et al., "Injection system of high density powder of cobalt, chromium, tungsten carbide". US patent 5996679 A, 7 November 1999.

74. "HIP-Cladding," Avure Technologies, [Online]. Available: http:// industry. avure. com. [Accessed 29 07 2014].

75. F. Czerwinski, "Thermochemical treatment of metals," in Heat Treatment-Conventional and Novel Applications Edited by F. Czerwinski, Rijeka, INTECH, 2012, pp. 73-112.

76. H. Du, M. Sommers and J. Agren, Metallurgical and Materials Transactions A, vol. 31, p. 2981, 2000.

77. K. Antymidis, G. Stergioudis, D. Roussos, P. Zinoviandis and D. Tsipas, "Boriding of ferrous and non-ferrous metals and alloys in fluidized bed reactor," Surface Engineering, vol. 28, pp. 255-259, 2002.

78. S. Timur et al, "Ultra-fast boriding of metal surfaces for improved properties". US Patent 2010/0018611 A1, 28 January 2010.

79. U. Sen, "Friction and wear properties of thermo-reactive diffusion coatings against titanium nitride coated steels," Materials and Design, vol. 26, no. 2, pp. 167-174, 2005.

80. T. Arai, "Tool treatment extends core and pin life in die casting operations," Die Casting Engineer, vol. March/April, 1999.

81. K. Strattford, C. Subramanian and T. Wilks, Surface Engineering vol. II: Engineering Applications, Cambridge, UK: Royal Society of Chemistry, 1993.

82. T. Ohmi and M. Iguchi, "Bonding strength of interafce between cast Mg-Al alloy and cast-in inserted transition metal cores," Journal of JSEM, vol. 13, pp. s189-s193, 2013.

83. "Magnesium Melting Crucibles," W. Pilling, Kesselfabrik GmbH & Co, KG, Altena, D-58762 Germany

84. "Magnesium die casting," [Online]. Available: http://www.dynacast. com. [Accessed 17 07 2014].

85. "Products for cold and hot chamber die casting of magnesium," [Online]. Available: www. omb-brondolin. com. [Accessed 5 07 2014].

86. "Materials datasheets," Exocor, 2014. [Online]. Available: www. exocor. com. [Accessed 2 09 2014].

87. Metals Handbook Vol 2, Materials Park, OH: ASM International, 1990.

88. "High performance steels for die casting," Kind & Co Edelstahlwerk, [Online]. Available: www. Kind-co. de. [Accessed 02 09 2014].

89. "PVD Coatings Application Guide," Chessen Group Inc., Mississauga, Ontario, 2000.

90. F. Lofaj et al., "Nanohardness and tribological properties of nc-TiB2 coatings," Journal of the European Ceramic Society, vol. 33, pp. 2347-2353, 2013.

91. "Certess SD Titanium Boron Nitride Coating," HEF USA, [Online]. Available: www. hefusa. Net. [Accessed 2 09 2014].

Citations

CHAPTER 1

I. Vyrides, E. Rakanta, T. Zafeiropoulou and G. Batis, "Efficiency of Amino Alcohols as Corrosion Inhibitors in Reinforced Concrete," Open Journal of Civil Engineering, Vol. 3 No. 2A, 2013, pp. 1-8. doi:10.4236/ojce.2013.32A001.

CHAPTER 2

Neha Patni, Shruti Agarwal, and Pallav Shah, "Greener Approach towards Corrosion Inhibition," Chinese Journal of Engineering, vol. 2013, Article ID 784186, 10 pages, 2013. doi:10.1155/2013/784186.

CHAPTER 3

Hualiang Huang, Guoan Zhang, Jiakuan Yang, Zhiquan Pan, and Xingpeng Guo, "Study of Flow-Assisted Corrosion of AZ91D Magnesium Alloy in Loop System Based on Array Electrode Technology," Journal of Chemistry, vol. 2015, Article ID 596740, 8 pages, 2015. doi:10.1155/2015/596740.

CHAPTER 4

E. Zacharopoulou, A. Zacharopoulou, A. Sayedalhosseini, G. Batis and S. Tsivilis, "Effect of Corrosion Inhibitors in Limestone Cement," Materials Sciences and Applications, Vol. 4 No. 12A, 2013, pp. 12-19. doi: 10.4236/msa.2013.412A003.

CHAPTER 5

Takakuwa, O. and Soyama, H. (2015) Effect of Residual Stress on the Corrosion Behavior of Austenitic Stainless Steel. Advances in Chemical Engineering and Science, 5, 62-71. doi: 10.4236/aces.2015.51007.

CHAPTER 6

Moreira, R. , Soares, T. and Ribeiro, J. (2014) Electrochemical Investigation of Corrosion on AISI 316 Stainless Steel and AISI 1010 Carbon Steel: Study of the Behaviour of Imidazole and Benzimidazole as Corrosion Inhibitors. Advances in Chemical Engineering and Science, 4, 503-514. doi: 10.4236/aces.2014.44052.

CHAPTER 7

Henry Hu, Xueyuan Nie and Yueyu Ma (2014). Corrosion and Surface Treatment of Magnesium Alloys, Magnesium Alloys - Properties in Solid and Liquid States, Dr. Frank Czerwinski (Ed.), ISBN: 978-953-51-1728-5, InTech, DOI: 10.5772/58929.

CHAPTER 8

Amany Fekry (2014). Corrosion Protection of Magnesium Alloys in Industrial Solutions, Magnesium Alloys - Properties in Solid and Liquid States, Dr. Frank Czerwinski (Ed.), ISBN: 978-953-51-1728-5, InTech, DOI: 10.5772/58942.

CHAPTER 9

Frank Czerwinski (2014). Corrosion of Materials in Liquid Magnesium Alloys and Its Prevention, Magnesium Alloys - Properties in Solid and Liquid States, Dr. Frank Czerwinski (Ed.), ISBN: 978-953-51-1728-5, InTech, DOI: 10.5772/59181.

Index

C

Chemical Vapour Deposition (CVD) 177
Coefficient of friction (COF) 185
Cold isostatic pressing (CIP) 260
Computational fluid dynamics (CFD) 49
Corrosion inhibitor 4, 5, 6, 8, 9, 10, 11, 12, 13, 14, 15, 17, 67, 76, 79, 82
Corrosion rate 230, 231
Corrosive environment 77
Counter electrode (CE) 52, 198

D

Diamond-Like Carbon Films (DLC) 178

E

Electrochemical impedance spectroscopy (EIS) 36, 195
Electrochemical measurement 1
Electrolytic plasma oxidation (EPO) 180, 182
Electrolytic Plasma oxidation (EPO) 180

F

Flow-assisted corrosion (FAC) 47, 48
Fresh water 34

H

Hot isostatic pressing (HIP) 260

Hypervelocity oxygen fuel spray (HVOF) 267

L

Linear Polarization Technique (LPR) 7
Liquid magnesium environment 229

M

Macro-segregation 235
Metal inert gas (MIG) 255
Mils per year (mpy) 141

P

physical vapour deposition (PVD) 178
Plasma electrolytic oxidation (PEO) 184
Plasma Electrolytic Oxidation (PEO) 184, 192
Plasma transferred arc (PTA) 255
Polyalthia longifolia (PL) 33

R

Rapid Chloride Permeability Test\" (RCPT) 69
Rare earth elements (RE) 158
Reference electrode (RE) 52

Resistor capacitor (RC) 207
Rotating cylinder electrode (RCE) 48
Rotating disk electrode (RDE) 48

S

Saturated calomel electrode (SCE) 6, 52, 199
Scanning electron microscope (SEM) 195
Scanning electron microscopy (SEM) 112
Stress Corrosion Cracking (SCC) 164

T

Tungsten inert gas (TIG) 255

W

Working electrodes (WE) 52
Working electrode (WE) 198

X

X-ray diffraction (XRD) 112

Z

Zero resistance ammeter (ZRA) 23